Inorganic Materials Chemistry

Mark T. Weller

University of Southampton

OXFORD   NEW YORK   TOKYO
OXFORD UNIVERSITY PRESS
1994

Oxford University Press, Walton Street, Oxford OX2 6DP

Oxford New York
Athens Auckland Bangkok Bombay
Calcutta Cape Town Dar es Salaam Delhi
Florence Hong Kong Istanbul Karachi
Kuala Lumpur Madras Madrid Melbourne
Mexico City Nairobi Paris Singapore
Taipei Tokyo Toronto
and associated companies in
Berlin Ibadan

Oxford is a trade mark of Oxford University Press

Published in the United States by
Oxford University Press Inc., New York

A catalogue record for this book is available form the British Library

Library of Congress Cataloging in Publication Data
(Data available)
ISBN 0 19 855799 X (Hbk)
ISBN 0 19 855798 1 (Pbk)

Printed in Great Britain by
The Bath Press, Avon

Series Editor's Foreword

Solid state chemistry has undergone a renaissance in this country in the last ten years. Synthetic methods, characterization techniques and underlying theories have now been developed that make a wider range of problems accessible to the chemist. Many of these deal with technologically important and intellectually demanding problems that make this an essential component of an undergraduate course.

Oxford Chemistry Primers are designed to give a concise introduction to all chemistry students by providing material that would usually be covered in an 8–10 lecture course. As well as giving up-to-date information, this series provides explanations and rationales that form the framework of an understanding of inorganic chemistry. Mark Weller presents the reader with an authoritative description of modern solid state chemistry, and the role of X-ray diffraction in it. Few if any standard inorganic textbooks can offer the kind of treatment that is achieveable by an expert in the field.

John Evans
Department of Chemistry, University of Southampton

Preface

The chemistry of inorganic materials is rapidly developing as an important component of university degree courses, reflecting recent advances at the forefront of research. Examples of such materials include the high-temperature superconductors, molecular sieves, fullerenes and layer intercalates. These topics also impinge on other undergraduate courses such as physics and materials science. However, the subject of inorganic solids is rarely covered in any depth in the major textbooks of inorganic chemistry. In addition the techniques used by the solid state chemist to prepare and characterize inorganic materials are seldom introduced alongside, or integrated with, the descriptive materials chemistry. This primer is intended to fill this omission by providing an affordable but comprehensive introduction to the chemistry of solid inorganic materials. The introductory chapters provide the foundation, reviewing basic crystallography and featuring a step-by-step guide to the analysis of powder diffraction data from simple systems. The synthesis of a wide range of inorganic material types is covered. These central techniques and concepts are developed in relation to fundamental solid state chemistry with discussion of simple oxide systems and a guide to the physical properties of solids. Finally the contemporary aspects of the subject are reviewed, integrating many of the ideas introduced in the preceding chapters.

I thank the members of my research group for their forbearance during the construction of this primer and Dr David Currie for useful suggestions and improvements. Finally I am indebted to three generations of my family for their constant support and abundant affection.

Southampton   M.T.W.
March 1994

Contents

1　Basic crystallography

1.1　Introduction

The structures of inorganic materials, the majority of which are crystalline in the solid state, are a key feature in the control of their chemistry and physical properties. The techniques used to identify and investigate such materials involve characterisation of the bulk structure. In order to discuss inorganic materials in some detail in the later chapters of this book it is important to cover the background to crystallography and structure determination.

1.2　Crystal systems and unit cells

All crystalline materials adopt, in the solid state, a regular distribution of atoms or ions in space. The simplest portion of the structure which is repeated by translation, and shows its full symmetry, is defined as the **unit cell**. In a two-dimensional array of ions, such as that shown in Fig. 1.1, the unit cell consists of a parallelogram. Any parallelogram may be chosen as a unit cell provided that translation along either of the cell directions repeats exactly the chosen unit.

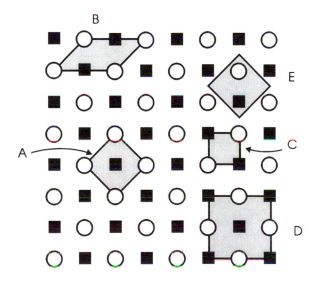

Fig. 1.1 Two-dimensional array demonstrating the choice of unit cell.

However, the unit cell is normally selected to be the simplest of these repeating units. Hence, in Fig. 1.1, the unit cell could be chosen as *A* for which displacements parallel to either edge of the square by the dimension of the unit cell produce a new position which is indistinguishable, in terms of

Note that it is possible, and equally valid, to choose a unit cell with the corners not coincident with the atoms, for example *E*.

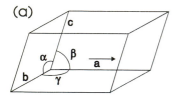

(a)

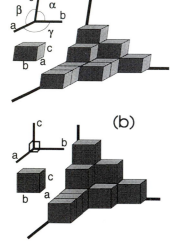

(b)

Fig. 1.2(a) General three-dimensional unit cell definition. (b) Packing of unit cells in three dimensions.

cell content and environment, from the original. The parallelogram B is also a suitable choice for the unit cell as it has the same area as A and shows the full translation symmetry. Square D would also be an acceptable choice of unit cell in terms of demonstrating the translational symmetry of the array, but is larger than A and B. However, parallelogram C is *not* a unit cell as translation parallel to one side by the length of the parallelogram places a corner originally at a ■ site on a ○ site; that is, the 'cell' does not show the translational symmetry of the ion array.

The basic unit cell in three dimensions is a parallelepiped with side lengths and angles defined as in Fig. 1.2(a). The angles and lengths used to define the size of the unit cell are known as the unit cell parameters; by convention the angle between a and b is γ, between b and c is α and between a and c is β. These parallelepipeds must stack together so as to form the structure and be completely space filling. That is, translation parallel to any of the sides by the length of that side will generate a new position for the unit cell; unit cells thus generated will fit perfectly together so that no part of the space is excluded. This is shown for two parallelepiped types in Fig. 1.2(b).

The unit cell shown as the parallelepiped in Fig. 1.2(a) has no symmetry in that the cell parameters and angles may take any values. An increasing level of symmetry produces relationships between the various cell parameters and leads to the seven **crystal classes**. These are summarised in Table 1.1 and sketched in Fig. 1.3.

A few conventions are worth noting when assigning particular lattice parameters for a particular crystal class. In monoclinic systems the non-right angle is normally taken as β; in the hexagonal and tetragonal systems the non-equivalent lattice parameter is normally taken as c. The trigonal systems can all be redefined using hexagonal lattices but this group is still generally regarded as a separate crystal system.

1.3 Fractional atomic coordinates and projections

The position of an atom within a unit cell is normally described using **fractional coordinates**. With respect to the unit cell origin, an atom within

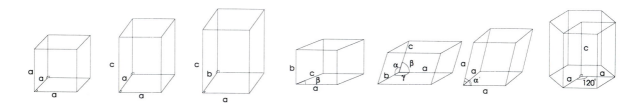

Fig. 1.3 Units cells of the seven crystal systems. From left to right, cubic, tetragonal, orthorhombic, monoclinic, triclinic, rhombohedral and hexagonal.

Table 1.1 The crystal systems

Unit cell dimensions		Crystal class	Example[†]
$a = b = c$	$\alpha = \beta = \gamma = 90°$	Cubic	NaCl, MgAl$_2$O$_4$, C$_{60}$K$_3$
$a = b \neq c$	$\alpha = \beta = \gamma = 90°$	Tetragonal	K$_2$NiF$_4$, TiO$_2$, BaTiO$_3$ (298 K)
$a \neq b \neq c$	$\alpha = \beta = \gamma = 90°$	Orthorhombic	YBa$_2$Cu$_3$O$_7$
$a \neq b \neq c$	$\alpha = \gamma = 90°\ \beta \neq 90°$	Monoclinic	KH$_2$PO$_4$
$a \neq b \neq c$	$\alpha \neq \beta \neq \gamma \neq 90°$	Triclinic	
$a = b \neq c$	$\alpha = \beta = 90°\ \gamma = 120°$	Hexagonal	LiNbO$_3$
$a = b = c$	$\alpha = \beta = \gamma \neq 90°$	Trigonal / Rhombohedral	BaTiO$_3$ below −80°C

[†] The majority of the example compounds are discussed later in the text.

the unit cell displaced by $x \times a$ parallel to a, $y \times b$ parallel to b and $z \times c$ parallel to c is denoted by the fractional coordinates (x,y,z). This method of describing the position of atoms within the unit cell is equally applicable to all crystal systems and cell sizes.

Three-dimensional representations of complex structures are frequently difficult to draw and to interpret and may be ambiguous. Often a clearer and faster method of representing three-dimensional structures is to draw the structure in projection viewing the unit cell down one direction, typically one of the unit cell axes. Normally the axis chosen is at 90° to the others and, hence, for triclinic systems a projection of the structure is, generally, of more limited use. The positions of the atoms relative to the projection plane are denoted by the fractional coordinate above the base plane. For example, the structure of zinc blende, ZnS, shown three-dimensionally in Fig. 1.4(a) may be viewed in projection down the z direction as in Fig. 1.4(b).

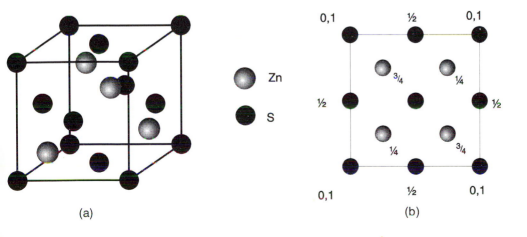

(a)   (b)

Fig. 1.4 Zinc blende structure viewed conventionally (a) and drawn as a projection (b).

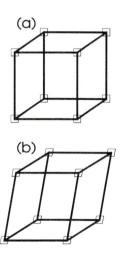

Fig. 1.5 A line of equally spaced points as a one dimensional lattice.

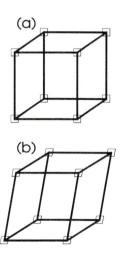

Fig. 1.6 Primitive lattices (a) cubic and (b) triclinic.

The environment of the additional atoms generated by the centring will be identical in all aspects to that of the original atom. Redefining the origin of the lattice at the face centred or body centred positions produces exactly the same arrangement of points.

1.4   Lattices

A **lattice** is defined as an array of equivalent points in one, two or, more normally for inorganic materials, three dimensions. The lattice provides no information on the actual positions of atoms or molecules in space but shows the translational symmetry of the material by locating equivalent positions. The environment of an atom placed on any one of these lattice points would be identical to that placed on any other lattice point. The simplest illustration of this is a one dimensional lattice consisting of an infinite series of equally spaced points along a line. The distribution of other lattice points about any randomly selected lattice point is identical, Fig. 1.5.

In three-dimensional crystal structures four lattice types are found. The simplest lattice type is known as **primitive**, given the symbol **P**, and a unit cell with a primitive lattice contains a single lattice point. That is, the only purely translational symmetry is that of the unit cell. Figs. 1.6a and b show examples of primitive lattices; lattice points are normally located at the corners of the unit cell parallelepiped

For the remaining three lattice types, as well as the translational symmetry of the unit cell, there is additional transitional symmetry *within* the unit cell. A second lattice type is body centred, which is given the symbol **I**, and an example of a body centred unit cell is shown in Fig. 1.7. This shows lattice points at the cell corners and, for body centring, the additional lattice point is at the cell centre with fractional coordinates $(\frac{1}{2},\frac{1}{2},\frac{1}{2})$. This means that if an atom is placed on a general position within a body centred unit cell with fractional coordinates (x,y,z) the lattice will generate a second, identical, atom at the position with fractional coordinates $(x+\frac{1}{2}, y+\frac{1}{2}, z+\frac{1}{2})$.

A three-dimensional lattice which has lattice points at the centre of *all* the unit cell faces, as well as at the corners, is known as face centred and given the symbol **F**, Fig. 1.8. The additional translational symmetry in this case consists of the three elements $+(\frac{1}{2},\frac{1}{2},0)$, $+(0,\frac{1}{2},\frac{1}{2})$ and $+(\frac{1}{2},0,\frac{1}{2})$, hence for an atom on a general site (x,y,z) three additional identical atoms will be generated within the unit cell with the coordinates $(x+\frac{1}{2},y+\frac{1}{2},z)$, $(x,y+\frac{1}{2}, z+\frac{1}{2})$ and $(x+\frac{1}{2},y,z+\frac{1}{2})$.

Finally a lattice which has points in just **one** of the faces is also known as face centred but given the symbol **A**, **B** or **C**, Fig. 1.9. A C type lattice refers to the case where the additional translational symmetry places lattice points at the centres of the faces delineated by the *a* and *b* directions as well as at the origin. In fractional coordinate terms, in addition to the general site (x,y,z) for a **C** centred lattice, a second site is generated at $(x+\frac{1}{2}, y+\frac{1}{2}, z)$. The **A** and **B** face-centred lattices are obtained in an identical manner but the additional lattice points occur in the *bc* and *ac* planes respectively. However, the **A** and **B** descriptions are not normally used, as redefinition of the *a*, *b* and *c* directions will produce the **C** centred description.

The lattice types describe the pure translational symmetry of the structure and show equivalent positions within unit cells related by translational symmetry. It should be noted that the lattice type does not provide definitive information on the number of atoms or molecules in the unit cell. In the case

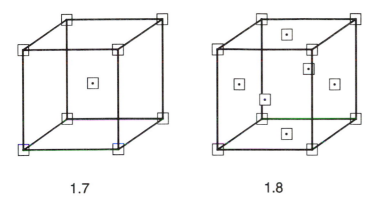

1.7   1.8   1.9

Figs. 1.7, 1.8 and 1.9 Body centred, I, face centred, F, and face centred, C, lattices respectively.

of a material crystallising with a primitive cubic lattice there may more than one atom or molecule of a particular kind within the unit cell, but their positions in terms of their environment will be quite different.

An example would be the unit cell of $YBa_2Cu_3O_7$, Fig. 8.5 which is primitive but contains two copper atoms in different environments, square planar and pyramidal.

1.5   Bravais lattices

The four lattice types (P,I,F,C) can be combined with the seven crystal classes, in Table 1.1, to give rise to all the possible variations, which are known as the **Bravais lattices**. These fourteen unique possibilities are summarised in Table 1.2 and it can be seen that not all combinations of lattice type with crystal system give rise to a different Bravais lattice. This is generally because an alternative description in terms of one of the fourteen

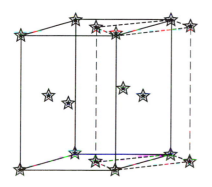

Fig 1.10 Equivalence of face centred and body centred tetragonal lattices.

(a)   (b)

Table 1.2 The fourteen Bravais lattices.

Crystal system	Bravais lattices
Cubic	P,I,F
Tetragonal	P,I
Orthorhombic	P,C,I,F
Monoclinic	P,C
Triclinic	P
Hexagonal	P
Trigonal/Rhombohedral	P(R)[‡]

[‡] The primitive description of the rhombohedral lattice is normally given the symbol R.

unique lattices is possible. For example in the case of a 'face centred tetragonal' unit cell, Fig. 1.10a, a new unit cell may be chosen which is body centred tetragonal, Fig. 1.10b.

1.6 Lattice planes and Miller indices

The lattice points which form an array in two or three dimensions showing the translation symmetry of the structure may be connected by **lattice lines** (two dimensions) or **lattice planes** (three dimensions). Each line or plane is a representative member of a *parallel set* of equally spaced lines or planes and each lattice point must lie on one of the lines or planes. Two possible lattice lines for a two-dimensional square lattice are shown in Fig. 1.11.

These lattice lines and planes are labelled using **Miller indices**. For a three-dimensional unit cell, three indices are required and designated conventionally h, k and l; the Miller indices for a particular family of planes are usually written (h,k,l) where h, k and l are integers, positive, negative or zero. In the case of a three-dimensional structure the derivation of Miller indices may be illustrated by considering the two adjacent lattice planes from the same set which cut the a, b and c axis at the origin and at the least distance along the cell directions. The Miller indices of this family of planes are given by the reciprocals of the fractional intercepts along each of the cell directions. For example, using the lattice plane shown in Fig. 1.12, the intercepts are at ½ × a, 1 × b and ⅓ × c. The Miller indices of this plane, one of a set of planes, are thus (2,1,3). The intercept along the unit cell a direction gives h, along b gives k and along c gives l. For planes which are parallel to one of the unit cell directions the intercept is at infinity and, therefore, the Miller index for this axis is $1/\infty = 0$. Fig. 1.13 shows an example of a (0,0,1) plane for a triclinic unit cell.

The separation of the planes is known as the d-spacing and is normally denoted d_{hkl}. From Fig. 1.12 it follows that this is also the perpendicular distance from the origin to the nearest plane. The relationship between d-spacing and the lattice parameter can be determined geometrically but is dependent upon the crystal system. In two dimensions simple trigonometry

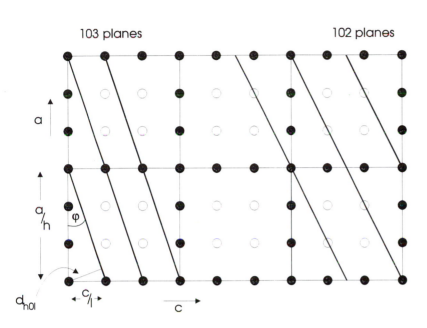

103 planes   102 planes

Table 1.3 Expressions for d-spacings in the different crystal systems.

Crystal System	Expression for d_{hkl} in terms of lattice parameters and Miller indices
Cubic	$$\frac{1}{d^2} = \frac{h^2 + k^2 + l^2}{a^2}$$
Tetragonal	$$\frac{1}{d^2} = \frac{h^2 + k^2}{a^2} + \frac{l^2}{c^2}$$
Orthorhombic	$$\frac{1}{d^2} = \frac{h^2}{a^2} + \frac{k^2}{b^2} + \frac{l^2}{c^2}$$
Hexagonal	$$\frac{1}{d^2} = \frac{4}{3}\left(\frac{h^2 + hk + k^2}{a^2}\right) + \frac{l^2}{c^2}$$
Monoclinic	$$\frac{1}{d^2} = \frac{1}{\sin^2\beta}\left(\frac{h^2}{a^2} + \frac{k^2 \sin^2\beta}{b^2} + \frac{l^2}{c^2} - \frac{2hl\cos\beta}{ac}\right)$$
Triclinic	Complex expression

Fig. 1.12 Definition of Miller indices in three dimensions. A pair of planes with Miller indices (213).

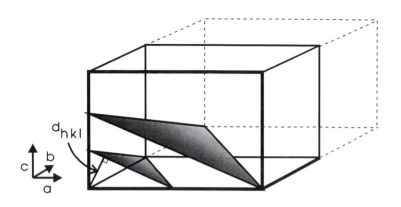

Fig. 1.12 Definition of Miller indices in three dimensions. A pair of planes with Miller indices (213).

may be used, with the definitions in Fig. 1.11, to derive the relationship between the Miller indices and d-spacing. Hence, from Fig. 1.11,

$$\sin\varphi = \frac{d_{h0l}}{(a/h)} \qquad \cos\varphi = \frac{d_{h0l}}{(c/l)} \qquad (1.1)$$

giving

$$d_{h0l}^2 = \frac{1}{\left(\dfrac{h}{a}\right)^2 + \left(\dfrac{l}{c}\right)^2} \qquad (1.2)$$

Extension of this expression to a three dimensional case with right angles between the lattice directions, an orthorhombic system, produces

$$\frac{1}{d^2} = \frac{h^2}{a^2} + \frac{k^2}{b^2} + \frac{l^2}{c^2} \qquad (1.3)$$

Fig. 1.13 (001) plane for a triclinic unit cell.

For the other crystal classes the relationships summarised in Table 1.3 apply.

1.7   X-Ray diffraction

X-Rays interact with electrons in matter and a beam of X-rays impinging on an inorganic material will be scattered in various directions by the atomic electrons. If the scattering centres are separated by distances comparable to the wavelength of the X-rays then interference between the X-rays scattered from particular electron centres can occur. For an ordered array of scattering centres this can give rise to interference maxima and minima.

While picometres are the SI unit that should be used in crystallography, angstroms are still

Distances between atoms or ions in solids are typically a few hundred picometres (equivalent to a few Ångstroms), which is comparable to the X-ray wavelengths produced by bombardment of metals by electrons. A study of the X-rays scattered by crystalline solids provides a wealth of structural information.

very widely employed; 1 Å = 10^{-10} m. This is because interatomic distances are conveniently expressed in angstroms, being typically between 1 and 4 Å. Both units are quoted where possible throughout the text.

Generation of X-rays

A beam of electrons striking a metal plate will, provided the electrons are sufficiently energetic, eject an electron from one of the metal atom core orbitals. Filling of this hole by electron decay from a higher energy orbital occurs with the emission of radiation. For moderately heavy metal atoms, for example first and second row transition metals, the transitions such as 3p→1s correspond to an emitted radiation wavelength in the range 0.5 to 3.0 Å (50-300 pm). In copper, for example, core electron vacancies formed by bombardment with electrons can be filled by decay from various higher energy electrons and thus the spectrum of X-rays obtained contains a number of intense maxima corresponding to the energies of these various transitions. In addition, X-rays can also be generated by the slowing down (or bremsstrahlung) of the electrons as they enter the metal target. A typical X-ray spectrum obtained from bombarding a metal target is shown in Fig. 1.14 with a background formed from the bremsstrahlung radiation and intense sharp maxima corresponding to the quantised electron transitions. The intense lines are labelled with respect to the electron shells which are involved in producing the X-ray. The filling of a vacant 1s orbital, principal quantum number one, is given a symbol K; this may be achieved, in a copper atom, by electron decay from the 2p or 3p levels and the X-ray produced by the transitions 2p → 1s and 3p → 1s are termed K_α and K_β respectively. Both these lines are in fact close doublets due to the spin multiplicity in the p shells; hence the K_α line ($\lambda \approx 1.54$ Å) in the copper spectrum consists of two lines at 1.5406 Å and 1.5444 Å termed $K_{\alpha 1}$ and $K_{\alpha 2}$. A transition which fills a 2s or 2p shell, principal quantum number 2, is given the symbol L and so on through the alphabet for higher principal quantum number shells.

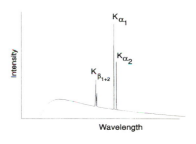

Fig. 1.14 The X-ray spectrum obtained from a copper target.

In order to carry out a diffraction experiment a single X-ray wavelength is desired. This may be achieved using a crystal monochromator in which the X-ray beam emerging from the X-ray tube impinges on a single crystal at a chosen, fixed orientation. For a particular angle θ and employing Bragg's Law (see the following section) only one wavelength can be diffracted from the crystal. By adjusting θ it is possible to select one wavelength from the X-ray spectrum, normally the most intense $K_{\alpha 1}$. A second way of removing unwanted wavelengths from the X-ray beam is to use a filter. This normally consists of a foil of a metal with an atomic number one or two below that of the metal used as the target to generate the X-ray spectrum. In the metal of the foil, transitions between energy levels will require slightly less energy than the corresponding transitions in the metal target due to a reduced nuclear charge and this leads to a high absorption coefficient for these X-rays. Hence, for an X-ray beam generated at a copper target with a nickel filter, X-rays with wavelengths less than ≈1.5 Å will be strongly absorbed. This will remove from the copper X-ray spectrum the majority of the bremsstrahlung radiation and, more importantly, lines such as the K_β. The X-ray beam emerging from the filter is, thus, almost monochromatic, consisting chiefly of K_α radiation. Both components of the K_α line are still present but they are so close in wavelength that diffraction experiments are possible producing close doublets in the diffraction pattern.

1.8   Scattering of X-rays by crystalline solids

Scattering of X-rays from crystalline solids can be demonstrated by consideration of the diffraction from points on a set of lattice planes such as those shown in Fig. 1.15.

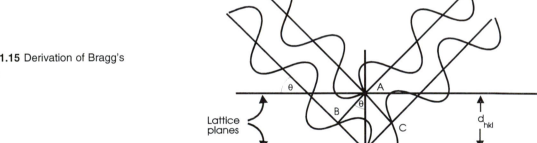

Fig. 1.15 Derivation of Bragg's Law.

The scattering of the impinging X-ray beam from the points A and D in neighbouring planes will produce *in phase* diffracted X-ray beams (constructive interference) if the additional distance travelled by the X-ray photon scattered from D is an integral number of wavelengths. This path difference BD + DC will depend on the lattice spacing or d_{hkl}, where *hkl* are the Miller indices for the planes under consideration, and will also be related to the angle of incidence of the X-ray beam, θ. For an integral wavelength pathlength difference the following relationship between θ and d_{hkl} can be obtained

$$\text{path difference} = \text{BD} + \text{DC} = 2.d_{hkl}.\sin\theta = n\lambda \qquad (1.4)$$

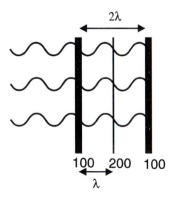

Fig. 1.16 Equivalence of $n = 2$ for the (100) reflection and $n = 1$ for the (200) reflection.

where *n* is an integer and λ is the X-ray wavelength. The expression $2d\sin\theta=n\lambda$ is known as the Bragg equation. In general *n* is always taken as unity; the reason for this is illustrated in Fig. 1.16. Rewriting the Bragg equation as $\lambda = 2.\,(d_{hkl}/n).\,\sin\theta$ and consideration of Fig. 1.16 shows that the higher order diffraction maxima, $n = 2$, is equivalent in representation to diffraction from a set of planes with half the separation. Hence, the second order diffraction from, for example, the 100 plane is identical to the first order diffraction from the 200 plane. In general, it is always possible to select lattice planes such that *n* is unity.

In a crystalline material an infinite number of lattice planes with different Miller indices exists and each set of planes will have a particular separation, dhkl. All the various d-spacings possible can be obtained by consideration of Table 1.3 for a particular crystal system. Each of the d-spacings generated for a set of lattice parameters and choice of integers for h,k and l can, in principle, through the Bragg equation give rise to a diffraction maximum at a particular diffraction angle, θ. The positions of these lines, in terms of the angle θ, can be obtained by combining the Bragg equation with the expression for the d-spacing for a particular crystal system and eliminating dhkl. For example, in the cubic system

$$\sin^2\theta \;=\; \frac{\lambda^2}{4a^2}\,\{\,h^2 \,+\, k^2 \,+\, l^2\,\} \qquad (1.5)$$

The scattering of an X-ray beam by a crystal will thus give rise to a large number of diffraction maxima, the positions of which are related to the lattice parameters, the Miller indices and the X-ray wavelength. Hence, by studying the diffraction of X-ray beams by crystalline solids structural information can be obtained.

The diffraction maxima in X-ray diffraction patterns are normally called reflections. Thus X-ray scattering from the 213 plane is termed the 213 reflection in diffraction pattern. This terminology whilst incorrect in terms of the process involved in origin of the maxima is so widely used that it is the accepted notation.

Note that slightly off the Bragg angle the diffracted X-ray beams from successive planes will only be *slightly* out of phase. However, in a crystal with a large number of planes, scattering from each successive plane will be further and further out of phase. Over the whole crystallite this leads to destructive interference.

1.9 The powder technique

In order for an X-ray beam, diffracted from a particular lattice plane in a crystalline sample, to be detected, the orientation of the X-ray source, crystal and detector must be correct. A powder or polycrystalline sample contains an enormous number of very small crystallites, typically 10^{-7} – 10^{-4} m in dimension and these particles will adopt, randomly, the whole range of possible orientations. An X-ray beam striking a polycrystalline sample will, therefore, be diffracted in all possible directions as governed by the Bragg equation. The effect of this is that each lattice spacing in the crystal will give rise to a cone of diffraction as shown in Fig. 1.17. In fact, each cone consists of a set of closely spaced dots each one of which represents a diffraction from a single crystallite within the powder sample. With a very large number of crystallites these dots join together to form the cone.

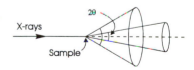

Fig. 1.17 Diffraction cones obtained from polycrystalline samples

Experimental methods
In order to obtain powder X-ray diffraction data in a form useful for analysis, the positions of the various diffraction cones need to be determined. This can be achieved by using either photographic film or a detector sensitive to X-ray

radiation. In both cases the basic idea is to determine the diffraction angle, 2θ, of the various diffraction cones.

The Debye–Scherrer camera

The Debye–Scherrer camera is the simplest technique used for acquiring powder diffraction data and is shown, schematically, in Fig. 1.18a. A strip of film is placed inside a cylindrical camera which has at its centre the sample. The X-ray beam enters through one side of the cylinder and is collimated on to the powder sample which is normally mounted by coating a glass fibre or filling a narrow glass capillary. The sample is often rotated around the axis of the fibre in order to present as many orientations of the crystallites as possible to the beam. The diffraction cones cut the film at various angles, 2θ. Once the film is developed and laid flat a typical powder diffraction pattern, as shown in Fig. 1.18b, is obtained. Each line of the film represents a lattice spacing in the sample; at low angles the lines have distinct curvature as the diffraction cone angle is small and a large part of the cone is intersected, at angles near 90° only a small section of the cone intersects the film and the line curvature is less marked. The position of the undiffracted beam can be seen as the dark area surrounding the shadow of the beam stop. From the radius of the camera and the distance along the film from the straight–through position, the diffraction angle, 2θ can be determined for each of the lines.

There is a trend for the diffraction maxima in the powder X-ray experiment to get weaker as 2θ increases. Strong reflections are, typically, seen up to $2\theta=50°$ but at high angles, e.g. above 90°, the reflections can be difficult to observe.

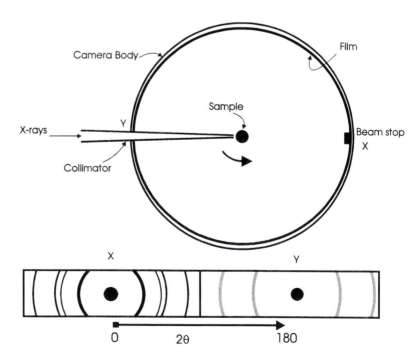

Fig. 1.18 The Debye–Scherrer camera and the form of the photographic diffraction pattern obtained.

One drawback of the Debye–Scherrer technique is that the defocusing of the X-ray beam once it has left the collimator results in relatively poorly resolved data, that is, the diffraction lines obtained on the film are quite broad. For simple structures which have relatively few lattice planes, and hence few lines in their diffraction patterns, this is of little consequence as their positions can still be readily obtained. However, for more complex structures with large unit cells and low symmetry the large number of different lattice d-spacings produce many diffraction cones. In such systems broad lines may overlap and it becomes difficult to measure their exact positions.

Other photographic techniques, principally the Guinier camera, use a crystal monochromator to focus the X-ray beam through the sample on to the film and this gives rise to much higher quality powder diffraction data.

The powder diffractometer

The powder diffractometer uses an X-ray detector, typically a Geiger–Muller tube or scintillation detector, to measure the positions of the diffracted beams. Scanning the detector around the sample along the circumference of a circle cuts through the diffraction cones at the various diffraction maxima, Fig. 1.19. The intensity of the X-rays detected as a function of the detector angle, 2θ, for a typical material is shown in Fig. 1.20. The reflection geometry frequently used for diffractometers is useful in that it is to some degree focusing, leading to reasonably well resolved data. This can be seen in Fig. 1.21, the sample acts like a mirror and provided the geometry of the equipment is correctly designed the diffracted beam can be focused on to the detector.

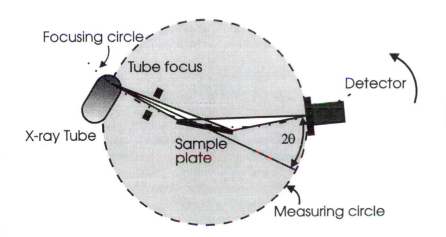

Fig. 1.19 Schematic of a typical powder diffractometer.

Fig. 1.20 A typical powder pattern obtained using a diffractometer.

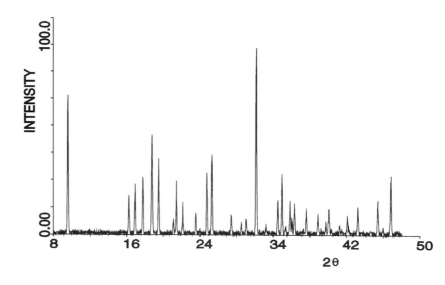

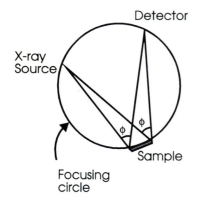

Fig. 1.21 Focusing in a powder diffractometer.

Comparison of the experimental techniques

The photographic methods have the advantage that very little sample is needed, typically a few milligrams, and the equipment is relatively cheap. Also, exposure times are reasonably short, generally <1 h, as all the diffracted X-rays in the different diffraction cones are collected simultaneously by the film. A disadvantage, in comparison with the diffractometer method, is that the film requires developing and analysis of the data can be time consuming; data collected on a diffractometer are readily stored in numerical form (2θ and counts) on a computer and analysis of this data in terms of determining peak positions is readily achieved. The other advantage of the diffractometer method is that intensities of the various reflections can be quickly ascertained; these can be extracted from photographic data using a densitometer but the task is time-consuming and less accurate.

1.10   Problems

1.1   What are the lattice types of NaCl (Fig. 4.1), UO_2 (Fig. 6.6), TiO_2 (Fig. 4.2) and K_2NiF_4 (Fig. 4.15).

1.2   Show that a body centred monoclinic lattice can be redrawn as, and is thus equivalent to, a C centred lattice.

1.3   Calculate the d spacings for 101 and -101 planes in an orthorhombic cell with dimensions $a = 3$ Å (300 pm), $b = 5$ Å (500 pm) and $c = 4$ Å (400 pm). Carry out the same calculation for the monoclinic cell with the same a, b and c but with ß=110°. Show diagrammatically, using a projection down c, the positions of these planes in the monoclinic system and, hence, explain why for this system $d_{110} \neq d_{-110}$.

2 The application and interpretation of powder X-ray diffraction data

2.1 General

The form of the powder X-ray diffraction data obtained from a material will depend upon the crystal structure it adopts. This structure is delineated by the lattice type, crystal class, unit cell parameters and the distribution of the various ion and molecule types within the unit cell. The number and positions, in terms of 2θ, of the reflections depend upon the cell parameters, crystal class, lattice type and wavelength used to collect the data, while peak intensity depends upon the types of atoms present and their positions.

As a result of the enormous range of different structures which materials adopt, nearly all crystalline solids have a unique powder X-ray diffraction pattern in terms of the *positions* of the observed reflections and the peak *intensities*. In mixtures of compounds each crystalline phase present will contribute to the powder diffraction pattern its own unique set of lines. The relative intensity of line sets from mixtures will depend upon the amount present and the ability of a structure to scatter X-rays.

The effectiveness of powder X-ray diffraction has led to it becoming the major technique used for the characterisation of polycrystalline, solid inorganic materials. Table 2.1 summarises some of the applications to which the technique is put and the most important of these are discussed in more detail in the following sections.

2.2 Identification of unknowns

Many of the powder diffraction data sets collected from inorganic, organometallic and organic compounds have been compiled into a data-base by the Joint Committee on Powder Diffraction Standards (JCPDS). This data base which contains over 50 000 unique powder X-ray diffraction patterns may be used to identify an unknown material from its powder pattern alone.

With simple materials this can be done by comparing two or three intense lines from the powder diffraction pattern with listings of the most intense lines in book form. For more complex materials a computer can be employed to look for coincidences between the experimental data and the data-base. For many solid materials this technique provides a rapid method of identification. Drawbacks are that the compound must have previously been made and its pattern entered in the data-base and that the compound must be crystalline.

Table 2.1 Applications of powder X-ray diffraction

Identification of unknown materials

Determination of sample purity

Determination and refinement of lattice parameters

Investigation of phase diagrams / new materials

Determination of crystallite size / stress

Structure refinement

Phase changes / expansion coefficients

2.3 Phase purity

In a mixture of compounds each crystalline phase present will contribute to the overall powder X-ray diffraction pattern. In preparative materials chemistry this may be used to identify the level of reaction and purity of the product. The reaction between the two solids Al_2O_3 and MgO to form $MgAl_2O_4$ may be monitored by powder X-ray diffraction, Fig. 2.1. At the start of the reaction a mixture of Al_2O_3 and MgO will produce an X-ray pattern

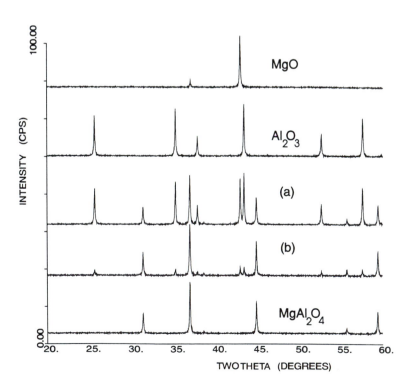

Fig. 2.1 The reaction between Al_2O_3 and MgO followed by powder X-ray diffraction.

combining those of the pure phases. As the reaction proceeds, Fig. 2.1 patterns (a) and (b), a new set of reflections corresponding to the product, $MgAl_2O_4$, emerges and grows in intensity at the expense of the reflections from Al_2O_3 and MgO. On completion of the reaction the powder diffraction pattern will be that of pure $MgAl_2O_4$. A materials chemist will often use powder X-ray diffraction to monitor the progress of a reaction in this way.

The powder X-ray diffraction method is widely employed to identify impurities in materials whether it be residual reactant in a product, as above, or an undesired by-product. However, the impurity must be crystalline, even a high level of amorphous phase would not be detectable except maybe as an increase in the background of the powder pattern. Also, the ability to detect a second, impurity phase will depend on how well the two components scatter X-rays though normally it is possible to detect a few per cent of an impurity phase. As X-rays are scattered by electrons, materials containing heavy elements, e.g. second and third row transition metals, diffract X-rays strongly and will produce, in general, strong reflections in the powder pattern. Materials containing light elements, e.g. organic compounds and first and second row oxides, will produce weaker patterns. As an example consider the reaction of Li_2O and Nb_2O_5 to form $LiNbO_3$. It would be easy to identify a few per cent of Nb_2O_5 as an impurity in $LiNbO_3$ but Li_2O would need to be present at a much higher level for its pattern to be observed in the powder diffraction data from a mixture of Li_2O, Nb_2O_5 and $LiNbO_3$.

Amorphous materials have no long range order but short range correlations produce broad diffraction maxima around positions corresponding to the interatomic distances.

2.4   Determination and refinement of lattice parameters

If Miller indices can be assigned to the various reflections in the powder pattern it becomes possible to determine the cell constants; these are directly related to 2θ and h,k,l through the relationships derived from the combination of Bragg's Law and the *d*-spacing expression and summarised in Table 1.3. This assignment may be readily achieved for cubic crystal systems with a simple relationship between the diffraction angles and lattice parameters. However, as the cell symmetry decreases, the number of lines in the diffraction pattern increases rapidly as the relationship between 2θ and the cell constants becomes more complex. In these cases indexing the powder diffraction data becomes a computational problem.

2.5   Indexing powder diffraction patterns from cubic systems

The relationship between diffraction angle and the lattice parameter is obtained by combining the Bragg equation with the expression for *d*-spacing in terms of lattice parameter and Miller indices. In the cubic crystal system the following expression was obtained in Section 1.8:

$$\sin^2\theta = \frac{\lambda^2}{4a^2}\{h^2 + k^2 + l^2\} \tag{2.1}$$

which can be written as

$$\sin^2\theta = A\{h^2 + k^2 + l^2\} \tag{2.2}$$

as both a and λ are constants. $\sin^2\theta$ values for the various reflections can be readily obtained from the diffraction angle 2θ. These will all be related to each other through a common multiplier A as h, k and l can only take integer values. Note that this will only be true for cubic systems; the more complex expressions for the d-spacings in terms of the lattice parameters in the other crystal systems means that more than one multiplier will be required. This argument may also be reversed in that *if* all the reflections in the powder diffraction pattern can be related through their $\sin^2\theta$ values by a single, common multiplier then the crystal system must be cubic.

The process involved in indexing data from a cubic system and obtaining the lattice parameter is, therefore, straightforward and is illustrated in Table 2.2. Once $\sin^2\theta$ values are tabulated they can be inspected for the common multiplier A; the ratio for each peak is the sum of the squares of the Miller indices.

Table 2.2 Indexing of the powder diffraction data from a cubic material

2θ	$\sin^2\theta$	Ratio	Miller indices
22.983	0.03969	1.00	100
32.729	0.07938	2.00	110
40.372	0.11907	3.00	111
46.962	0.15876	4.00	200
52.908	0.19845	5.00	210
58.418	0.23814	6.00	211
68.595	0.31752	8.00	220

Note that there are some values which $h^2+k^2+l^2$ cannot take. For instance it is impossible to chose the integer values of the Miller indices such that the sum of their squares equals 7 or 15. This means that there can be no peak in the powder diffraction pattern corresponding to $7A$ or $15A$.

Once the data have been indexed, derivation of the lattice parameter is readily undertaken. Any peak may be chosen and the Miller indices, wavelength and $\sin^2\theta$ value for that peak substituted back into eqn 2.1 to give a. For example, from Table 2.2 where the data were collected with $\lambda = 154$ pm

$$0.31752 = 154^2.8/4a^2$$

$$a = 386.5 \text{ pm} \quad (3.865 \text{ Å})$$

In general, if all lines have been measured with the same precision of 2θ then the use of a high angle line will produce a more accurate lattice parameter. In practice, a least squares refinement technique using all the reflection positions is used by powder diffractionists to obtain the best value for the lattice parameter.

2.6   Tetragonal and hexagonal systems

Combination of the expression for the *d*-spacings in a tetragonal unit cell with the Bragg equation and replacement of constant terms by *A* and *C* gives the following expressions

tetragonal $\sin^2\theta \;=\; A\,(\,h^2 + k^2\,) + Cl^2$ (2.3)

hexagonal $\sin^2\theta \;=\; A\,(\,h^2 + hk + k^2\,) + Cl^2$ (2.4)

where *A* and *C* incorporate the wavelength and the lattice parameters. For these crystal systems $\sin^2\theta$ values obtained from the powder diffraction will be related through two multipliers. By consideration of the possible values which the Miller indices can take it is feasible to derive the relationships between the various $\sin^2\theta$ values. This is done for a tetragonal system in Table 2.3.

Obviously, assigning these ratios to particular peaks can be quite difficult. However, the method can be illustrated for a tetragonal system by consideration of a system where there is only a slight distortion from a cubic unit cell; that is, the lattice parameters *a* and *c* are very similar. In this case the relationship between the reflections seen in the powder diffraction pattern of a tetragonal material to those of a cubic phase is shown schematically in Fig. 2.2. The 100 reflection of a cubic material is made up of diffraction from the (100), (010), (001), (−100), (0−10) and (00−1) lattice planes which all have an identical *d*-spacing. The number of planes which contribute towards a reflection is termed the multiplicity, in this case the multiplicity is 6. The $\sin^2\theta$ values for all these reflections are identical as they depend only on $(h^2+k^2+l^2)$. In a tetragonal material with slightly different *a* and *c* lattice parameters the constants *A* and *C* will be very similar. Hence, the $\sin^2\theta$ values for the (100) and (001) reflections will diverge only marginally and these maxima will be observed at only slightly different 2θ values in the powder diffraction pattern. In Fig. 2.2 the peak doublet near 22° can be assigned as deriving from the 100 reflection of the cubic material and these two reflections indexed as 001 and 100. Lattice planes contributing to the (100) peak will be the (100), (010), (−100) and (0−10) giving a multiplicity of 4, those contributing to the (001) reflection will be the (001) and (00−1) giving a multiplicity of 2. The multiplicity is reflected in the peak intensities which have a 2:1 ratio, Fig. 2.2.

In this slightly distorted material, assignment of Miller indices to all the other reflections has now become easy. Once the 100 and 001 reflections

Table 2.3 $\sin^2\theta$ relationships in the tetragonal system, $h,k,l \leq 2$

A
C
2A
A+C
2A+C
4A
4C
4A+C
A+4C
5A
5A+C
4A+4C

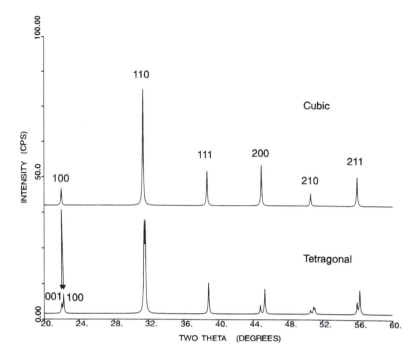

Fig. 2.2 Comparison of the powder X-ray diffraction patterns of cubic and slightly tetragonally distorted BaTiO₃.

have been indexed, values of the constants A and C can be obtained using eqn. 2.3 and the other observed $\sin^2\theta$ values assigned using the possible ratios shown in Table 2.3. Again, following full indexing of the data, the values of A and C may be used to calculate the a and c parameters. For the compound in Fig. 2.2, BaTiO₃, the values obtained are 399.5 pm (3.995 Å) and 403.4 pm (4.034 Å). BaTiO₃ is discussed more fully in Chapter 4.

2.7   Lattice type and systematic absences

In indexing the powder patterns it has been assumed that all the possible reflections are observed, that is, scattering from each of the different lattice planes is sufficiently intense to contribute to the diffraction profile. This is normally so for a primitive lattice, but for body centred and face centred lattices restrictions occur on the values that h, k and l may take if the reflections are to have any intensity. This results in certain reflections not being observed in the powder diffraction pattern and these are known as systematic absences. The origin of these absences can be illustrated with regard to Fig. 2.3. Consideration of diffraction from the 100 plane of a cubic material with a body centred lattice shows that X-rays diffracted at the Bragg angle from the 100 planes (the faces of the cube) will be, by definition, in phase. However, halfway between the 100 planes, as a result of the body centring, there is an identical plane of atoms shifted in the x and y directions by (½,½). X-Ray diffraction from this plane, the 200, will have a pathlength difference relative to that from the 100 planes exactly $\lambda/2$ out of phase and will, therefore, destructively interfere with it. In a body centred

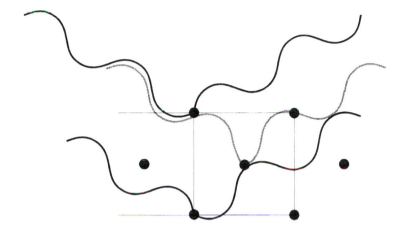

Fig. 2.3 Origin of 100 absence in a body centred lattice. The scattering from the lattice point at the cell centre is 180° out of phase with that from the cell corners.

lattice there is one lattice point in the 200 plane for every lattice point in the 100 plane and overall there will be total destructive interference for the 100 reflection. Note that it does not matter where the atoms or ions actually are in the unit cell, it is the fact that they always occur in pairs (at (x,y,z) and at $(x+\frac{1}{2}, y+\frac{1}{2}, z+\frac{1}{2})$) generated by the body centring that causes the destructive interference.

Extension of these considerations to other possible values of h, k and l in a body centred structure show that for a reflection to be *present* then $h+k+l$ is even. Alternatively this may be expressed as the condition, $h+k+l=2n+1$ is the systematic absence for I lattices. Similar considerations for a face centred cubic lattice show the following restrictions on h, k and l for a reflection to be observed: $h+k=2n$, $k+l=2n$ and $h+l=2n$. This restriction may also be expressed in the terms, h, k and l must all be odd or all even. Table 2.4 summarises which reflections are observed for the various lattice types in cubic systems.

Using this information it is possible, once a data set has been indexed, to determine the lattice type from the systematic absences. The existence of these absences can be a complicating factor when attempting to index cubic systems. If a material adopts a face centred cubic lattice then the common ratio between the $\sin^2\theta$ values may not be immediately obvious. For this F lattice the first few peaks, if all are observed, will have $\sin^2\theta$ values in the ratio $1.00 : 1.333 : 2.666 : 3.666 : 4.00$. However, there is still a single common multiplier giving the integral ratios $3:4:8:11:12$ for the first few reflections corresponding to the (111), (200), (220), (311) and (222) reflections.

In a body centred cubic system the systematic absences can also be misleading in terms of indexing. The $\sin^2\theta$ values of the first few lines will be

Table 2.4 Reflection conditions imposed by lattice type

			Lattice type		
Miller indices			Primitive, P no systematic absence	Body centred, I $h+k+l=2n$	Face centred, F h,k,l all odd or all even
1	0	0	✓	×	×
1	1	0	✓	✓	×
1	1	1	✓	×	✓
2	0	0	✓	✓	✓
2	1	0	✓	×	×
2	1	1	✓	✓	×
2	2	0	✓	✓	✓
2 2 1 / 3 0 0			✓	×	×
3	1	0	✓	✓	×
3	1	1	✓	×	✓
2	2	2	✓	✓	✓
3	2	0	✓	×	×
3	2	1	✓	✓	×
4	0	0	✓	✓	✓

found to have the ratios 1:2:3:4:5:6:7:8 corresponding to the ratios of the sums of the squares of the Miller indices. The presence of a line with a ratio of seven, points to these lines originating from a body centred system as there are no values of h, k and l which square and sum together to give seven. The common multiplier must be adjusted, by being halved, for these lines to give the ratios 2:4:6:8:10:12:14:16 corresponding to the Miller indices (110), (200), (211), etc. These are the first reflections found for a material adopting a body centred cubic structure.

Other symmetry elements in crystal structures can also give rise to absences in the powder diffraction pattern and sometimes peaks are not observed experimentally, as their intensity is so low that they cannot be distinguished from the background noise. One case where peaks have such a low intensity as to give rise to apparent absences in the powder diffraction pattern can be illustrated by consideration of KCl. X-Rays are scattered by electrons and isoelectronic species such as K^+ and Cl^- will scatter X-rays to almost identical degrees. Therefore, in the face centred cubic sodium chloride structure adopted by KCl, Fig. 2.4, the ion positions all scatter X-rays to an almost equivalent degree, that is, the potassium and chloride ions are practically indistinguishable to X-rays. Hence, the lattice repeat *observed by*

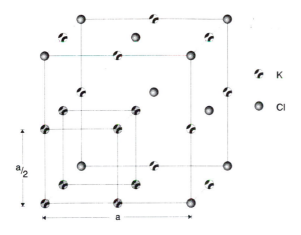

Fig. 2.4 The structure of KCl showing the primitive unit cell of isoelectronic species.

X-rays is much smaller than the true one and is the smaller cube outlined in Fig. 2.4. This unit cell is, if the corner atoms are considered identical as they are to X-rays, primitive and the powder X-ray diffraction pattern of KCl can be indexed using a primitive cubic unit cell of dimensions half that of the true face centred cubic unit cell. In essence the scattering equivalence of K^+ and Cl^- has made the reflections with odd Miller indices from the face centred cubic structure so weak that they are not observed. It is then possible to re-index the peaks with even Miller indices (by halving them) using a smaller common ratio which corresponds to the smaller (half size) unit cell, Fig. 2.5. The fact that the powder pattern of KCl, and other similar compounds with isoelectronic species, for example MgO, can be indexed using a primitive unit cell is merely a consequence of the origin of the X-ray scattering which makes such species indistinguishable.

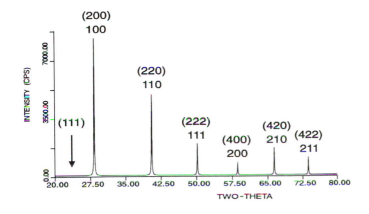

Fig. 2.5 Powder X-ray diffraction pattern from KCl. Indices in parentheses refer to the true unit cell while those directly above the peaks are from the 'primitive' subcell. The position of the zero intensity (111) reflection from the true unit cell is also marked.

2.8 Peak intensities

The peak intensities of the various reflections in an X-ray diffraction experiment are controlled by the atom types and their distribution in the unit cell. Measurement of these intensities allows the crystallographer to work out the structure of the crystalline material which produced them. This structure determination is, normally, carried out using the *single crystal X-ray diffraction* technique. In a single crystal experiment the diffraction data is collected using a detector or film covering two dimensions and the orientation of the crystal being studied can also be adjusted. Using this technique the intensities of all the individual reflections may be obtained and employed to determine the full crystal structure.

In the powder X-ray diffraction experiment, because all the reflections occur along the single 2θ axis, reflections frequently overlap, particularly at high 2θ values. Extracting intensities for the individual reflections, essential for the structure determination, becomes very difficult. Structural information may still be obtained from powder diffraction patterns using a technique known as the Rietveld method. In this method, a trial structure (in which likely positions, in the unit cell, are assigned to the different atoms present) is used to calculate intensities for the various reflections. These intensities may then be combined with the various factors controlling the form of the powder pattern, e.g. lattice parameters, radiation wavelength, to generate a 'calculated' powder pattern based on the trial structure. Comparison of this profile with the experimental one can then be undertaken and adjustment of the trial structure parameters carried out in order to obtain the best fit between the experimental and calculated data. This then provides reasonably accurate information on the atom positions in the unit cell. However, as stated above, a trial structure is required before the technique can be employed.

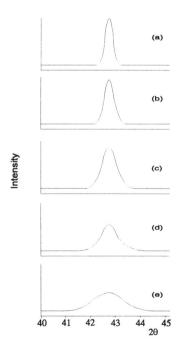

Fig. 2.6 The effect of crystallite size on peak width in powder X-ray diffraction pattern: (a) instrumental broadening, (b) 1 μm particles, (c) 100 nm, (d) 10 nm, (e) 5 nm.

2.9   Other applications of powder X-ray diffraction

Crystallite size

In order to observe sharp diffraction maxima in the powder X-ray diffraction pattern, the crystallites need to be of sufficient size to ensure that slightly away from the 2θ maximum, destructive interference occurs. The requirement is for a sufficiently large number of planes that summation of the diffracted waves, which are only slightly out of phase between successive planes, eventually produces destructive interference. In materials with very small crystallites, with few planes, the diffracted X-ray intensity slightly away from the Bragg angle does not totally destructively interfere and this leads to broadening of the reflections, Fig. 2.6.

The Scherrer formula relates the thickness of the crystallites to the breadth of peaks in the powder X-ray diffraction pattern:

$$t = \frac{0.9\lambda}{\sqrt{B_M^2 - B_S^2}\ \cos\theta} \tag{2.5}$$

where t is the crystallite thickness, λ the X-ray wavelength, θ the Bragg angle (half the measured diffraction angle) and B_M and B_S are the width in radians of diffraction peaks of the sample and a standard at half height. The standard peak width is obtained from a highly crystalline sample (with a diffraction peak at a similar diffraction angle to that of the sample under investigation) and represents the instrumental broadening.

The Scherrer method of determining crystallite size from powder data is of wide application. For materials which are of plate-like habit, for example clays, the size of the crystallites in the different directions may be ascertained by consideration of the peak widths of the various $(h00)$, $(0k0)$ and $(00l)$ reflections.

Variable temperature studies

Combination of powder X-ray diffraction equipment with a furnace or cryostat allows powder diffraction patterns to be collected over a large temperature range. Studying the evolution of the pattern as a function of temperature allows rapid identification of phase changes, which cause marked changes in the pattern as the material adopts a new structure, and the determination of thermal expansion coefficients. For the latter, the peak positions are monitored as a function of temperature, allowing determination of the lattice parameters and unit cell volume.

2.10   Problems

2.1 The following powder X-ray diffraction data (λ=154 pm (1.54 Å)) were obtained from zeolite 4A:

2θ (°)   7.144   10.121   12.405   14.302   16.005   17.563   20.300

Index the data and determine the lattice parameter and type. The structure of zeolite 4A is shown in Fig. 7.4a; estimate the size of the main channel.

2.2 Plutonium metal undergoes a phase transition when heated to 750 K changing from a face centred cubic structure with a = 463.7 pm (4.637 Å) to a body centred cubic structure with a = 363.8 pm (3.638 Å). Determine the peak positions expected in the powder X-ray diffraction pattern (λ =154 pm) of plutonium metal above and below 750 K in the 2θ range 0–75°. How would the expansion of plutonium metal on heating above 750K affect the form of the diffraction pattern?

2.3 Determine the expected appearance of the 310 reflection of cubic $BaTiO_3$ once the compound has transformed to the tetragonal phase.

3 The synthesis of inorganic materials

The synthesis methods used in the preparation of inorganic materials, many of which have extended lattices rather than discrete molecules, are quite different from those used by organic, organometallic and coordination chemists. Rather than altering a single functional group or ligand attached to a molecule, the materials chemist works with complete lattices, either building or modifying them. The techniques used to do this and their application to inorganic materials synthesis are discussed in this chapter.

3.1 High temperature reactions

The most widely used method for the synthesis of inorganic materials follows an almost universal route that involves heating the components together at high temperature over an extended period. Many of the inorganic materials discussed in this book are complex oxides, i.e. materials containing more than one metal in addition to oxygen. These compounds include ternary phases such as $BaTiO_3$, and quaternary oxides, e.g. $YBa_2Cu_3O_{7-x}$. In such cases the direct high temperature reaction of the component oxides frequently yields the desired complex oxide phase. This direct method of preparation is, however, equally applicable to other inorganic material types, for example the following syntheses

Chlorides $\qquad$ $3CsCl + 2ScCl_3 \rightarrow Cs_3Sc_2Cl_9$ $\qquad$ (3.1)

Aluminosilicates $\qquad$ $NaAlO_2 + SiO_2 \rightarrow NaAlSiO_4$ $\qquad$ (3.2)

3.2 The process of the reaction between solids

In order to illustrate a typical synthesis in solid state chemistry the preparation of the ternary oxide, $SrTiO_3$, will be used as an example. Simple binary oxides are generally available commercially as polycrystalline powders with typical particle dimensions of a few micrometres. In the synthesis of $SrTiO_3$, SrO and TiO_2 would be ground together in the correct stoichiometric proportions (1:1 molar ratio) using a pestle and mortar. A representative reaction would then involve pressing the mixture into a pellet, transferring it to a crucible and placing this in a furnace at 900°C. The crucible is normally constructed of an inert material such as vitreous silica, recrystallised alumina or platinum.

A schematic view of the reaction mixture, magnified many times, is represented in Fig 3.1a and consists of particles of the constituent oxides, separated mainly by voids, but with some contact at particular crystal faces. Reaction between the binary oxides occurs by migration of ions between the pure oxides across these interfaces and formation of the new structure of the target ternary oxide. A representation of a partially reacted mixture, derived from that in Fig. 3.1a is shown in Fig. 3.1b. In reactions between oxides the most mobile ions are normally the cations, as these are usually smaller than the oxide ions, and it is these ions that diffuse through the oxide and between the reactant blocks.

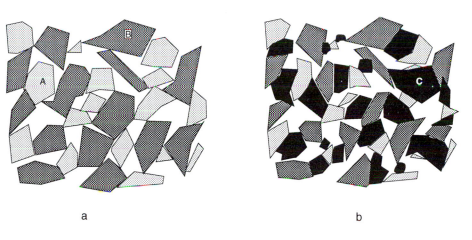

a   b

Fig. 3.1 Schematic of reacting particles in a solid state reaction.

The need for the high temperatures required in the synthesis of mixed oxides can be rationalised by consideration of the energetics of the reaction process. In comparison with organic and organometallic chemistry where a typical reaction will involve the breaking of *one* bond and the formation another, sometimes simultaneously, the solid state reaction involves a complete disruption of the structure. The cations in an oxide material are normally coordinated to a large number of oxide ions ranging from 4 for the smallest ions, e.g. Li^+, to 12, or more for ions such as barium. In order for these ions to migrate to the interfaces/grain boundaries and form a new structure, considerable energy is required to overcome the lattice energy. Hence, the typical reaction temperatures of oxide chemistry in the range 500–2000°C. At the lowest temperatures, ions with a low charge to radius ratio, for example Cs^+ and other alkali metals, which have the weakest interaction with the lattice, will undergo reaction. So, for example, the reaction

$$Na_2O_2 \ + \ 2CuO \rightarrow 2NaCuO_2 \tag{3.3}$$

proceeds reasonably rapidly at 400°C. Small, strongly charged cations, which interact more strongly with the oxide lattice, require higher temperatures to produce significant reaction. For example the reaction

$$MgO \;+\; Al_2O_3 \;\rightarrow\; MgAl_2O_4 \tag{3.4}$$

would normally be carried out at temperatures above 1400°C. Increasing the temperature of a solid state reaction will speed up the process as the rates of diffusion of the various ions increase. The limit of this behaviour is seen when the mixture of solids melts and the ions become highly mobile. Occasionally the reaction to form a new complex oxide may be carried out in the melt; however, as discussed in Section 3.6, crystallisation of a molten reaction mixture will often not give the desired product.

Besides raising the temperature, other methods can be used to enhance the reaction rate between solids. Consideration of the schematic reaction mixture in Fig. 3.1a helps illustrate these. To improve the reaction rate, the number of interfaces at which the reaction can occur needs to be maximised. This may be achieved by both reducing the particle size and pelletising the material. Thorough grinding of the metal oxide mixture will produce a homogenous distribution of the two oxides ensuring that each oxide particle is surrounded, as far as possible, by oxide particles of the other type. Pelletising the material, normally carried out using a hydraulic press, reduces the inter-particle void space (typically to less than 10%) producing many more interfaces over which the reaction can take place.

After the reaction has proceeded for some time a volume of the product builds up along the interface between the reacting crystallites. Further formation of the ternary phase requires the cations to migrate over longer distances, though this distance is minimised by reducing the particle size in the initial grinding. To facilitate further reaction the partially reacted powder is normally reground at intervals, this has the advantage of introducing pristine reactant crystal faces to each other.

3.3   Precursor, solution and gel methods

The reduction in the particle size by grinding the reaction mixture, even mechanically, can only achieve a limiting level of grain diameter, at best about 0.1 μm. However, chemical methods can be used to effectively reduce particle size still further and so enhance reaction rates at a particular temperature. One valuable method is the use of precursors that decompose at moderate temperatures to generate the reactant oxides; such precursors include metal carbonates, hydroxides and nitrates. When the crystallites of these precursors disintegrate with the loss of gaseous species they break down to yield fine, very reactive particles resulting in faster reaction times. An additional advantage of precursors is that they are normally air stable while many oxides are hygroscopic and/or pick up carbon dioxide from the air. A commercial sample of BaO normally contains $Ba(OH)_2$ and $BaCO_3$ and will continue to react with water/CO_2 in the air whilst being weighed out in the laboratory. This makes the stoichiometry of a reaction involving BaO difficult to control. $BaCO_3$, however, is perfectly air stable, and is frequently employed for reactions involving barium, though it needs to be heated strongly to ensure complete decomposition to the oxide.

In order to produce very homogeneous mixtures the solid state chemist can employ solutions where the reactant ions are dispersed on the atomic scale. Precipitation from mixed ion solutions will produce a solid containing the required ions, though care must be taken to ensure that the precipitate is of the correct stoichiometry. A typical reaction would be

$$Zn^{2+}(aq) + 2Fe^{2+}(aq) + 3C_2O_4^{2-}(aq) \rightarrow ZnFe_2(C_2O_4)_3 \downarrow \qquad (3.5)$$

where the highly insoluble oxalates are precipitated from a 1:2 molar aqueous solution of zinc and iron. Heating the mixed metal oxalate product in air above 800°C results in its decomposition to yield the ternary oxide, $ZnFe_2O_4$, in just a few hours.

Precipitation methods have considerable advantages, it is, however, difficult to ensure that the precipitate has the desired stoichiometry. In the preparation of complex oxides, such as $YBa_2Cu_3O_7$, the correct conditions, for example pH, to precipitate a 1:2:3 Y:Ba:Cu complex from solution are difficult to determine and one ion may be left, to some degree, in solution leading to an impure product. This may be overcome using gel methods. In this technique rather than precipitating and filtering to produce a solid, the whole solution is solidified through complexation and/or removal of water producing a gel. Such techniques have been widely used for the synthesis of ceramics including the high temperature superconductor, $YBa_2Cu_3O_7$. Dissolution of yttrium, barium and copper as their nitrate salts in water is followed by addition of citric acid and ethanediol. Heating of this mixture and evaporation of water results in the formation of a gel that may then be heated further in a crucible to form the complex oxide. The final stage, due to the mixing of various ions at the atomic level, can be carried out at temperatures as low as 700°C and requires only a few hours. This should be contrasted with a direct reaction of oxides which typically requires temperatures above 900°C for several days.

3.4 Sealed tubes and special atmospheres

The direct reaction discussed above is widely used for the synthesis of complex oxides, fluorides, chlorides, phosphates, silicates and sulphides, and is generally carried out in air. However, the reaction environment may need to be controlled if a particular oxidation state or reaction stoichiometry is targeted. In such cases the reactions are carried out in a controlled atmosphere, using a tube furnace, where a particular gas can be passed over the reaction mixture during heating. An example of such a reaction would be the use of an inert gas to prevent oxidation, as in the preparation of the thallium(I) compound $TlTaO_3$

$$Tl_2O + Ta_2O_5 \xrightarrow{N_2 / 600°C} 2TlTaO_3 \quad\quad (3.6)$$

Another example would be the reduction in hydrogen of V_2O_5 to VO_2

$$V_2O_5 \xrightarrow{H_2 / 1000°C} VO_2 \quad\quad (3.7)$$

High gas pressures may also be used to control the reaction product stoichiometry. For example the use of oxygen under high pressures, several hundred atmospheres, permits the formation of Sr_2FeO_4 (Fe(IV)) from mixtures of SrO and Fe_2O_3. This should be contrasted with the formation of compounds containing Fe(III) under ambient pressures.

Another problem encountered in direct synthesis is the volatility of one or more components of the mixture at the high temperatures required for reaction. In such cases the reaction mixture is normally sealed in a glass tube, under vacuum, prior to heating. Examples of such reactions are

$$Ta + S_2 \xrightarrow{500°C} TaS_2 \quad\quad (3.8)$$

$$Tl_2O_3 + 2BaO + 3CaO + 4CuO \xrightarrow{860°C} Tl_2Ba_2Ca_3Cu_4O_{12} \quad (3.9)$$

where the sulphur and thallium(III) oxide, respectively, are volatile at the reaction temperature and would be lost from the reaction mixture in an open vessel, leading to impure products.

3.5   Solution and hydrothermal methods

Many non-oxide inorganic materials may be synthesised by crystallisation from solution. Whilst the methods are very diverse, the range can be illustrated by the following;

$$3KF(aq) + MnCl_2(aq) \rightarrow KMnF_3\downarrow + 2KCl(aq) \quad\quad (3.10)$$

$$ZrO_2 + 2H_3PO_4 \rightarrow Zr(HPO_4)_2.H_2O + H_2O \quad\quad (3.11)$$

$$12NaAlO_2 + 12Na_2SiO_3.9H_2O \xrightarrow{90°C} Na_{12}[Si_{12}Al_{12}O_{48}].nH_2O \text{ (Zeolite 4A)}$$
$$+ 24NaOH \quad\quad (3.12)$$

Fig. 3.2 The tetrapropyl-ammonium ion directing the channel structure of ZSM-5.

The application of solution methods is extended by using hydrothermal techniques where the reacting solution is heated in a sealed vessel above its boiling point. Such reactions are important in the synthesis of open structure aluminosilicates (zeolites). The preparation of zeolites typically involves crystallisation from a strongly basic aqueous solution containing the tetrahedral structure building ions ($NaAlO_2$ and SiO_2) plus a templating ion that is frequently an organic base. The shape of the templating ion directs the crystallisation of the aluminate and silicate tetrahedra and determines the structure of the zeolite product. The crystallisation process is frequently slow and hydrothermal methods act to speed up the process. An example of zeolite preparation is that of ZSM-5, a material discussed in more detail in Chapter 7. A mixture of silicic acid ($SiO_2.nH_2O$), sodium hydroxide, aluminium sulphate, water, *n*-propylamine and tetrapropylammonium bromide is heated in a hydrothermal bomb at 160°C for several days. The organic template ion directs the formation of the structure as shown in Fig. 3.2 with the alkyl groups filling the channels. The template ion may be removed from the cavity by heating the product in air, oxidising the organic cation and leaving intact the aluminosilicate framework.

3.6   Phase diagrams and synthesis

The direct reaction of components to form inorganic materials is generally carried out below the melting point of the system despite the process being extremely slow. Ideally the materials chemist would like to use molten phases where reaction rates are much faster due to the more rapid ion diffusion. However, cooling a molten mixture does not normally yield the desired product in a pure form and to understand this behaviour a knowledge of phase diagrams is required.

Phase diagrams are very useful to the solid state chemist; they illustrate the behaviour of multi-component systems as a function of temperature by delineating the ranges of stability of various phases in terms of the stoichiometry of the mixture and temperature.

A simple two component system phase diagram is shown in Fig. 3.3 with the *x* axis denoting the relative proportions of two components A and B and the *y* axis the temperature. A and B do not react together, i.e. there is no compound A_aB_b in the system; an example of this behaviour would be CuO and CdO. The diagram has four regions. At low temperatures both A and B are solids and a mixture of the two consists of discrete particles of the two solids. At higher temperatures A and B melt and, depending on the stoichiometry of the mixture temperature/composition, regions occur where solid A exists with a liquid, solid B exists with a liquid or the whole system is molten. The composition/temperature point E, i.e. the stoichiometry which is totally liquid at the *lowest* temperature, is known as the eutectic point and this composition is the eutectic composition. The curve separating totally liquid from solid+liquid regions is known as the liquidus.

The above case is of little interest to the synthetic chemist as no compound is formed in the system. However, the simple eutectic system behaviour is found in more complex phase diagrams such as that shown in Fig. 3.4. In this case the components A and B react together and form a compound of stoichiometry AB with a different structure. This new compound is represented in the figure by a vertical line. A feature of this type of phase

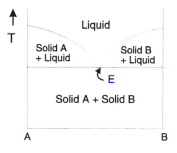

Fig. 3.3 Simple binary eutectic system.

diagram is that if the compound AB is heated then on reaching the temperature T_1 the compound melts to produce a liquid of the same composition. This is known as *congruent* melting. For this type of phase diagram the overall behaviour is that of two simple, binary, eutectic phase diagrams joined side to side. Hence, the left-hand side of the diagram is a binary eutectic between A and AB and the right-hand side one between AB and B. In a binary system that adopts this type of phase diagram the synthesis of AB from the components A and B can be undertaken very easily. A 1:1 stoichiometric mixture of A and B is melted; on cooling the melt, if fully homogeneous, will crystallise to give pure AB.

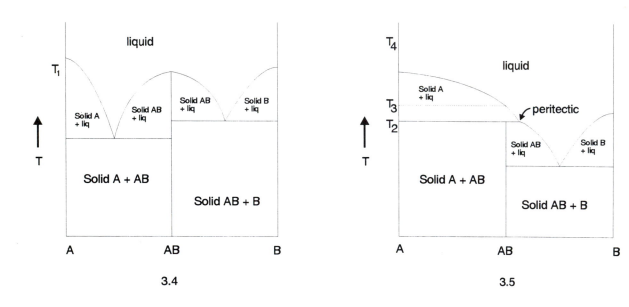

3.4

3.5

Figs. 3.4 and 3.5 Two component phase diagrams forming a phase AB with congruent and incongruent melting respectively.

The other type of phase diagram that occurs for a two component system forming a compound AB is shown in Fig. 3.5, where the behaviour of the compound AB on heating is somewhat different. In this case when the compound AB reaches the temperature T_2 it *partially* melts to yield some solid A plus a liquid that will be richer in B. On further heating the solid A dissolves in the liquid until at temperature T_3 the system is completely molten and the liquid is of the composition AB. This behaviour, where AB does not melt directly to give a liquid of the same composition, but deposits one component, generating a liquid richer in the other is known as *incongruent* melting and has profound effects regarding synthesis. In this system if a molten 1:1 mixture of A and B is cooled to temperature T_3, solid A will crystallise out and the liquid will get richer in B, its composition following the liquidus curve. On reaching the temperature T_2 the compound AB is now stable and the reaction

$$A(s) + B(l) \rightarrow AB(s) \qquad\qquad (3.13)$$

can occur. However, this reaction involves a solid and, as has been described above, will be very slow, particularly so as solid A will have settled out on the bottom of the reaction vessel. In addition as the particles of A start to react back with the liquid, to form solid AB, the product will coat the solid A inhibiting further reaction. If the system is cooled extremely slowly, typically over many days then the reaction between solid A and liquid B will occur and pure AB can be obtained. However, faster cooling causes the liquid rich in B to solidify before the reaction with solid A is complete yielding a mixture of AB and B plus some solid A which crystallised out initially.

Hence, for this type of phase diagram, with incongruent melting, heating the system above T_2, known as the peritectic temperature, the maximum temperature at which the compound AB is stable, is unhelpful. In such systems the preparation of AB is normally carried out below T_2 where AB is stable but both A and B are solid, that is, the typical, direct reaction of solids discussed above in Sections 3.1 and 3.2.

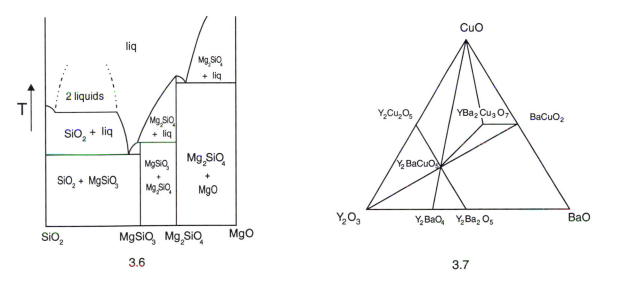

Figs. 3.6 and 3.7 The MgO / SiO_2 phase diagram and the three component system CuO–BaO–Y_2O_3 at 900°C respectively.

More complex binary phase diagrams with several compounds of differing stoichiometry, for example Fig. 3.6, are constructed from regions with congruently and incongruently melting phases. In this figure Mg_2SiO_4 melts congruently whilst $MgSiO_3$ melts incongruently. An understanding of the simpler binary phase diagrams discussed above is all that is required to interpret this and even more complex looking diagrams.

In systems with more than two components the complexity of the phase diagram increases. For three components triangular diagrams are used to represent the phases occurring and the temperature axis is not shown, for example, Fig. 3.7. The information shown on the diagram represents a *single temperature*; as this changes then the phases that are stable and shown on the diagram may alter. Each line across the diagram between phases would represent a binary phase diagram of the type discussed above.

3.7   Low temperature methods

The synthesis methods described above are applicable to most solid state reactions where a new structure is built from the reactants. This generation of a new ion or bonding arrangement requires considerable energy, demanding the high reaction temperatures common in solid state chemistry. However, some reactions of solids can be carried out at much lower temperatures, at or just above room temperature, and involve *modification* of a material's structure. These reactions include intercalation or insertion where an ion or molecule is added to a compound leaving the basic structure of the solid unchanged and ion-exchange when one ion within a material is replaced by another.

Fig. 3.8 Schematic of an intercalation reaction.

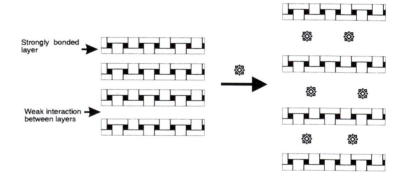

Intercalation in layer structures

Many materials that undergo intercalation reactions have layer structures of the type shown schematically in Fig. 3.8. Such compounds have strongly bonded two dimensional layers which only weakly interact with the other layers above and below, typically through van der Waals type forces. The simplest compound of this type is graphite shown in Fig. 3.9 in which hexagonal rings of carbon form infinite layers separated by 350 pm. Adjacent layers are displaced relative to each other in the layer direction so that the repeat distance along the c direction, perpendicular to the layers, is twice the layer separation. The weak van der Waals interactions between the layers mean that ions and molecules may be readily incorporated or intercalated between the layers. For example, the reaction

$$8C(s) + K(l) \rightarrow C_8K \tag{3.14}$$

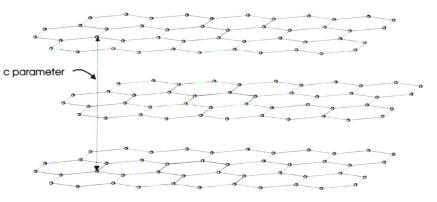

Fig. 3.9 The layer structure of graphite.

Table 3.1 Some graphitic intercalation compounds

Compound	Intercalate stoichiometry
Br_2	C_8Br, $C_{28}Br$
H_2SO_4	$C_{96}HSO_4$, $C_{48}HSO_4$, $C_{24}HSO_4$
K	C_8K, $C_{24}K$, $C_{36}K$, $C_{48}K$
$FeCl_3$	$C_{12}FeCl_3$, C_7FeCl_3
OsF_6	C_8OsF_6

takes place above the melting point of potassium, 64°C, with the driving force for the reaction being electron transfer from the potassium to the graphite layers. A better description of the product is $C_8^-K^+$. The potassium ions occupy sites between the graphite layers as shown in Fig. 3.10. As the carbon–carbon bonds in the layers remain intact throughout the reaction, the energy required for this process is much less than typical values for reactions involving solids, hence the low reaction temperature. Potassium is just one example of an ion or molecule which can be inserted into graphite and a selection of species is given in Table 3.1.

Other materials that adopt crystal structures consisting of weakly interacting layers undergo similar reactions to those of graphite. Examples include transition metal sulphides, such as TaS_2 and VPS_3, and FeOCl. The transition metal disulphides have been found to intercalate some large and complex species; examples include long chain alkyl amines $C_nH_{2n+1}NH_2$ up to $n = 18$ and organometallics such as cobaltocene. The structure and properties of these layer compound intercalates are discussed in more detail in Chapter 7.

Insertion compounds of metal oxides

Insertion reactions involving large molecules normally require a layer type structure in order for sufficient room to be generated, by increased separation of the layers, to accommodate the guest species. Very small ions, e.g. Li^+,

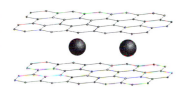

Fig. 3.10 The structure of C_8K.

may be inserted into three-dimensional structures, such as oxides, provided there is sufficient room. In the case of alkali metal insertion into transition metal oxides this process occurs with reduction of the transition metal. One example is the insertion of lithium into rhenium trioxide via the reaction

$$0.3LiI(s) + ReO_3(s) \rightarrow Li_{0.3}ReO_3(s) + 0.15I_2(s) \qquad (3.15)$$

The structure of ReO_3 is shown in Fig. 4.3.

This reaction may be carried out by grinding the two solids, anhydrous lithium iodide and rhenium trioxide, in a pestle and mortar. The rhenium trioxide structure is sufficiently open to allow rapid diffusion of lithium ions to empty sites and the metal oxide framework remains intact throughout the reaction. The rhenium is partially reduced from Re(VI) to Re(V). Other transition metal oxides including WO_3, MoO_3, V_2O_5 undergo similar insertion reactions with lithium and other small ions such as Na^+ and H^+.

Ion exchange

Structures of the type described above, with strongly bonded layers or frameworks containing weakly held guest molecules or ions, frequently undergo facile reactions at low temperatures in which one weakly bonded species is swapped for another. This ion exchange behaviour of zeolites is well known. For example, the process

$$\{2Na^+/zeolite\ 4A\} + Ca^{2+}\ (aq) \rightarrow \{Ca^{2+}/zeolite\ 4A\} + 2Na^+(aq) \qquad (3.16)$$

where sodium ions in a zeolite framework are replaced by calcium ions from solution, is widely used in water softening processes.

Similar reactions can occur for other materials that have ions weakly interacting with a strongly bonded framework. One example is $LiTiNbO_5$ which has a layer structure of the type schematically shown in Fig. 3.11. The lithium ions may be exchanged with other ions such as sodium and potassium by stirring the solid with a solution of the desired cation.

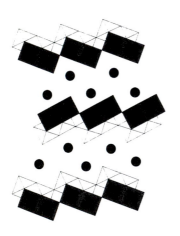

Fig. 3.11 The structure of $LiTiNbO_5$ with linked octahedra (NbO_6 and TiO_6) forming layers separated by Li ions.

3.8   Problems

3.1 The reaction $MgO + Al_2O_3$ only occurs at a reasonable rate at temperatures above 1400°C. How could $MgAl_2O_4$ be prepared at lower temperatures?

3.2 A compound of the stoichiometry $KNbTiO_5$ may be prepared via two methods (i) reaction of K_2CO_3, Nb_2O_5 and TiO_2 at high temperature and (ii) reaction of Li_2CO_3, Nb_2O_5 and TiO_2 at high temperature followed by stirring the product with potassium nitrate solution at 50°C. The products have different structures. Explain.

3.3 Explain what happens if a 40:60 mixture of A and B is cooled from T_4 in a system adopting the phase diagram shown in Fig. 3.5.

4   Transition metal oxides

4.1   Review of binary transition metal oxide structures

MO

The first row transition metal monoxides TiO to NiO adopt the rocksalt
structure shown in Fig. 4.1. This would be predicted from the ionic radius
ratio, r_{M2+}/r_{O2-}, between 0.414 and 0.732, and filling of the octahedral site in
a close-packed array of oxide ions. Many of these materials are non-
stoichiometric existing over a range of composition and this behaviour is
described in detail in Chapter 6. The electronic properties of these compounds
which range from metal to insulator are discussed in Chapter 5. CuO has a
structure formed from linked CuO_4 square planes, a more typical geometry for
Cu^{2+} than the octahedral sites found in the rocksalt structure. Few second and
third row transition metals form monoxides.

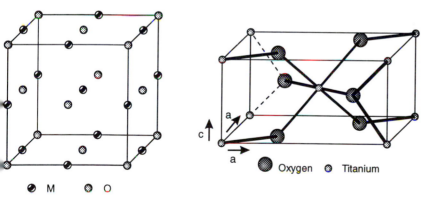

Oxygen   Titanium

M   O

Figs. 4.1 and 4.2 The rocksalt structure and the rutile structure respectively.

MO_2

The majority of transition metal dioxides crystallise with the rutile structure,
shown in Fig. 4.2, though in some materials the structure is slightly distorted.
Materials of this stoichiometry exist for Ti, Cr, V and Mn in the first row
transition metal series and for Zr–Pd and Hf–Pt in the second and third rows.
The rutile structure is again predicted for the majority of transition metal
dioxides on the basis of radius ratio rules and filling of sites in a close-packed
oxide structure.

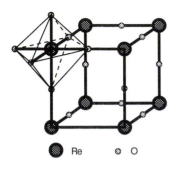

Fig. 4.3 The ReO$_3$ Structure. One of the ReO$_6$ octahedra is outlined.

The rutile unit cell is tetragonal with the metal atoms lying on the cell corners and body centre; however, because the oxide ions are not related through the body centred condition the unit cell is primitive. The co-ordination geometry around the metal ion is approximately octahedral, but four metal to oxide ion distances are, in general, slightly different from the other two.

MO$_3$

Few transition metals adopt the +6 oxidation state in oxide and, therefore, oxides of the stoichiometry MO$_3$ are rare. ReO$_3$ has the structure shown in Fig. 4.3 which may be easily constructed by linking ReO$_6$ octahedra through all vertices. This gives rise to a primitive, cubic unit cell. WO$_3$ adopts a slightly distorted variant of the ReO$_3$ structure, but still contains solely corner sharing octahedra.

4.2   Ternary oxides

The structure of many binary oxides can be predicted on the basis of the relative sizes of the metal and oxide ions and filling of holes in a close packed oxide lattice. Such predictions of structure are more difficult for ternary phases. The combination of two or more metals in an oxide generates a wealth of structural possibilities dependent on the relative sizes of the two metal ions and the oxide ion. In addition the stoichiometry of the ternary oxide may be changed by varying the proportions of the two component oxides and, for transition and lanthanide elements, the oxidation state. For example, at least twenty ternary oxide phases are formed between strontium and vanadium including SrV$_2$O$_6$, Sr$_2$V$_2$O$_5$, SrVO$_3$ and Sr$_2$VO$_4$. The structural chemistry of ternary and more complex oxides is, thus, an extensive subject. There are, however, a few structures which are widely adopted by ternary oxides and many materials from these structural classes have important technological applications. These ternary oxide structures are also used as units in building more complex oxides, such as the high temperature superconductors, Chapter 8.

4.3   The perovskite structure

The perovskite structure, frequently adopted by materials of the stoichiometry ABO$_3$ is, perhaps, the most widespread ternary phase. Perovskites are named after the mineral CaTiO$_3$ which was identified by the Russian mineralogist L.A. Perovski. Several hundred materials of stoichiometry ABO$_3$ adopt the perovskite structure or a slighted distorted version. In the notation ABO$_3$ the A cation is conventionally the larger of two metal ions so, for example, in SrTiO$_3$ [r(Sr^{2+})= 1.52 Å and r(Ti^{4+}) =0.745 Å)] strontium would be denoted as the A cation and titanium as the B cation.

The perovskite structure is shown in Figs 4.4a and 4.4b with the A and B cations at the unit cell origin respectively. The coordination geometry of the smaller B cation is octahedral while that of the larger A cation is 12 fold.

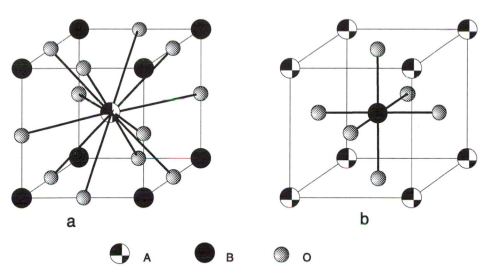

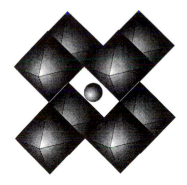

Fig. 4.4 The perovskite structure with the origin at the B (a) and A (b) cations.

The relationship of the structure to that of ReO_3 is readily appreciated by reference to Figs. 4.5 and 4.3. The BO_6 octahedra link all vertices forming the same cubic framework as in ReO_3. The addition of the A cation at the centre of the unit cell produces the ABO_3 perovskite stoichiometry.

The range of oxides adopting the perovskite structure is delineated by the relative sizes of the cations A and B, provided that their overall charge sums to six. The dimensions of the sites accommodating the A and B cations mean that the perovskite structure is frequently found for ternary ABO_3 oxides formed with one large and one small cation. Some examples are given in Table 4.1

In some complex oxides the perovskite sites may be shared by two or more cations. A complicated looking formula such as La_2MgRuO_6 may be rewritten as $La(Ru_{1/2}Mg_{1/2})O_3$, which is a perovskite with La on the A site and a random mixture of Ru and Mg on the B sites.

Fig. 4.5 The perovskite structure using linked octahedra.

Table 4.1 Perovskite cation types

Charge on A	Charge on B	Example
3+	3+	$LaCrO_3$
2+	4+	$SrTiO_3$
1+	5+	$NaWO_3$

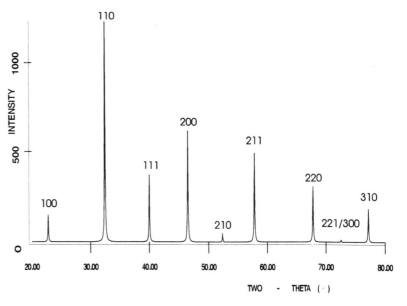

Fig. 4.6 The powder X-ray diffraction pattern of $SrTiO_3$.

The cubic perovskite unit cell has a primitive lattice and, therefore, gives rise to the typical powder X-ray diffraction pattern with no general absence. Fig. 4.6 shows the powder diffraction pattern obtained from $SrTiO_3$.

Tolerance factors in perovskites

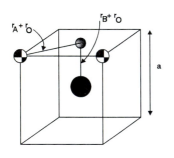

Fig. 4.7 The relationship of ionic radii in perovskites.

For ternary and more complex oxides, including perovskites, the size and coordination preferences of three or more ions need to be satisfied simultaneously by the structure; it is, however, rarely possible that all these preferences can be accommodated perfectly. This can be illustrated for the perovskite structure by deriving a relationship between the radii of the various ions. Consideration of Fig. 4.7 yields the expressions

$$a = 2 \times B\text{-}O = 2 \, (r_B + r_O) \qquad (4.1)$$

and

$$a = \frac{1}{\sqrt{2}} \times 2 \times A\text{-}O = \sqrt{2} \times (r_A + r_O) \qquad (4.2)$$

where a is the unit cell parameter and r_A, r_B and r_O are the ionic radii of A, B and O respectively. Hence, for the perovskite structure to provide the *ideal* contact distance for both the A and B cations

$$2 \times (r_B + r_O) = \sqrt{2} \times (r_A + r_O) \qquad (4.3)$$

However, ionic radii are not fixed values, but figures derived from consideration of the coordination preferences of a particular ion in a wide range of crystal structures. Some variation in the ion separations from $(r_+ + r_-)$ can thus be countenanced in any crystal structure and this can be built into the above relationship by introduction of a tolerance factor, t.

$$2 \times t \times (r_B + r_O) = \sqrt{2} \times (r_A + r_O) \qquad (4.4)$$

For a perfect fit for the two ions into the perovskite structure $t = 1$; increasing deviation of t from unity indicates increasing strain on the perovskite structure in terms of the ions being too big or too small to fit on their allotted site. Ionic radii normally used in the above expression are those of Goldschmidt as these provide the best predictive power in terms of defining the stability of the perovskite structure for different cation sizes. However, other systems of ionic radii can be used to illustrate how well the two cations fit into the structure and, for consistency, those of Shannon and Prewitt used elsewhere in this primer will be employed. Taking the values for Sr^{2+} ($r = 158$ pm), Ti^{4+} ($r = 74.5$ pm) and O^{2-} ($r = 126$ pm) yields a value of 1.002 for t indicating that the coordination distance preferences for both strontium and titanium are met. As the values for t depart from unity increased stresses are built up in the perovskite structure as one or both of the cations occupies a site of non-ideal dimensions. This may be overcome by a distortion of the perovskite structure and for t values in the range $0.85 < t < 1.06$ a perovskite structure, or a distorted version thereof, is frequently adopted by compounds of the stoichiometry ABO_3. Using the Shannon and Prewitt radii, for $0.9 < t < 1.0$ the structure is often cubic, above 1.0 and below 0.9 the structure is normally distorted. For t values outside the range 0.85–1.06 the perovskite structure cannot meet the coordination requirements of the two cations and other structures are normally adopted. For example, $LiNbO_3$ and $FeTiO_3$ have $t \ll 0.9$, as the cations are of similar sizes, and the perovskite structure is not formed; the structure of $LiNbO_3$ is described in Section 4.5.

Note that these ranges only apply to Shannon and Prewitt radii: for other systems, e.g. Goldschmidt, the perovskite stability ranges for t are quite different.

BaTiO$_3$

For $BaTiO_3$ a tolerance factor of 1.06 is obtained; this value, greater than unity, indicates that the titanium ion is occupying a site that is larger than it would prefer. Above 120°C $BaTiO_3$ adopts a perfect cubic perovskite structure; at these high temperatures the extensive thermal motion of the various ions makes packing considerations in terms of ionic radii less important. On cooling, $BaTiO_3$ undergoes a series of phase changes as summarised in Table 4.2.

Table 4.2 Phase behaviour of BaTiO$_3$

Transition temperature		−80°C		5°C		120°C	
Crystal Structure	Rhombohedral		Orthorhombic		Tetragonal a = 3.995 Å c = 4.034 Å		Cubic a = 4.002 Å

These phase changes are a result of the structure distorting from the perfect perovskite description so as to develop a more acceptable co-ordination geometry around the titanium atom. This is shown in projection

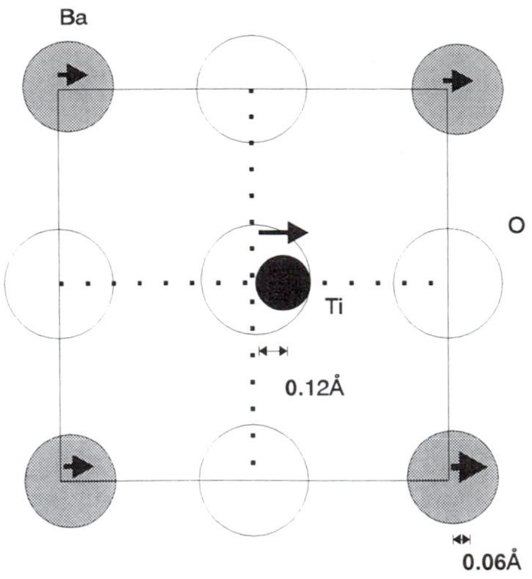

Ba

Ti

O

0.12Å

0.06Å

Fig. 4.8 Cation displacements relative to the oxide sublattice in tetragonal $BaTiO_3$.

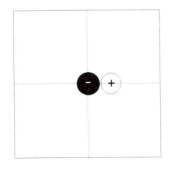

Fig. 4.9 The centres of positive and negative charge in the tetragonal unit cell of $BaTiO_3$.

Centrosymmetric structures have an inversion centre at the origin, that is for every atom on a site x,y,z in the unit cell there is an equivalent atom at $-x,-y,-z$. In ferroelectric crystals, therefore, no inversion centre is present leading to a relatively low crystal symmetry.

in Fig. 4.8 with the barium ions at the unit cell corners. In the tetragonal phase the major positional shift is that the titanium ion becomes displaced from the central site towards one of the oxide ions in the cell faces; smaller displacements of the barium and oxygen ions also occur. These displacements make the lattice directions inequivalent such that $a=b\neq c$, that is the cell symmetry reduces to tetragonal, though the distortion is quite small with c/a = 1.01. However, this distortion is readily observed in the powder diffraction pattern, Fig. 2.2. The indexation of the powder pattern from $BaTiO_3$ was carried out in Section 2.6.

One effect of this displacement of titanium from the central site of $BaTiO_3$ is on the charge distribution in the unit cell. If charges are formally attached to the various ions as Ba^{2+}, Ti^{4+} and O^{2-} then the centres of negative and positive charge in the cubic form of $BaTiO_3$ coincide at the cell centre. However, in the tetragonal phase, owing to the large displacement of titanium ion, the unit cell charge centres are no longer coincident, Fig. 4.9. This polarisation of charge gives the unit cell a high dipole moment.

The displacements of the titanium ions in the large number of unit cells which make up part of a crystal or *domain* of tetragonal $BaTiO_3$ can be arranged such that in neighbouring unit cells the displacements are in exactly the same directions. The result of this is that the individual unit cell dipoles all line up in the same direction, Fig. 4.10a, and this behaviour means that tetragonal $BaTiO_3$ is termed a ferroelectric. Ferroelectric crystals are non-centrosymmetric and display a long range ordering of the dipoles below a characteristic temperature known as the Curie temperature. In the absence of an external electric field, the ordering of the dipoles within the various domains (sections of the crystal) produces a spontaneous polarisation of the crystallite. Following the application of an external electric field the direction of polarisation in the various domains can order so that they all line up with the external field. The polarisation of the crystal results in a very high dielectric constant for $BaTiO_3$ and it is widely used in capacitors.

Charge separation in distorted perovskite structures is a common feature though the interaction of ion displacements in neighbouring unit cells may occur in a manner different from that observed in $BaTiO_3$. In the tetragonal phase of $PbZrO_3$ the lead ion displacements in adjacent unit cells are in opposite directions, Fig. 4.10b, and this material is an antiferroelectric.

$BaTiO_3$, in common with a number of other ferroelectrics shows what is known as piezoelectric behaviour. A piezoelectric material develops an electric polarisation on the application of stress. Compression of a crystal of $BaTiO_3$ causes further distortion of the tetragonal $BaTiO_3$ unit cell, a further separation of the charges and a higher degree of polarisation of the charge. This may be used in a number of ways to convert mechanical energy into electrical energy; examples include applications as record player needles and in gas lighters.

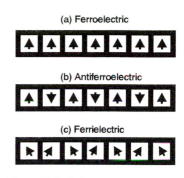

Fig. 4.10 Ordering schemes for dipole moments.

4.4   Insertion compounds and perovskite structures

In Section 3.7 the reaction of LiI with ReO_3 to give $Li_{0.3}ReO_3$ was discussed. The perovskite and ReO_3 structure are identical except for the filling of the A sites in the former. The reason for the ease of this reaction can now be readily seen in that the structure of the product, $Li_{0.3}ReO_3$, is a perovskite with a proportion, 30%, of the A sites filled. Owing to the small size of Li^+ the perovskite structure distorts slightly by tilting of the ReO_6 octahedra so as to bring oxygen ions closer to it. The insertion of lithium ions into ReO_3 merely involves diffusion of Li^+ into the ReO_3 structure through the faces of the unit cell cube. The Re–O bonds forming the framework are unaffected allowing this solid state reaction to progress at room temperature. The powder X-ray diffraction patterns of $Li_{0.3}ReO_3$ and ReO_3 are almost identical, as they are dominated by scattering from the rhenium (due to the high number of electrons), except for a small shift in peak positions caused by a decrease in lattice parameter.

Tungsten trioxide is another material which undergoes insertion reactions of this type. WO_3, with a slightly distorted version of the ReO_3 structure, forms the compound $Na_{1.0}WO_3$, adopting the perovskite structure. $H_{0.6}WO_3$ synthesised by inserting protons into WO_3, has a structure of the perovskite type, but the hydrogen atoms, instead of sitting isolated on the perovskite A sites, form OH groups with the oxide ions along the unit cell edges.

4.5   LiNbO$_3$

Lithium niobate whilst having the ABO_3 stoichiometry does not adopt the perovskite structure. The ionic radii of Li^+ and Nb^{5+} are too similar to fit the perovskite structure, producing a very small tolerance ratio; in addition lithium normally occupies sites with low coordination numbers, 4 or 6. The structure adopted by $LiNbO_3$ is shown in Fig. 4.11 where the lithium ions are coordinated to six oxygens. The lithium ions may be displaced within the octahedral sites and $LiNbO_3$ is a ferroelectric. $LiNbO_3$ is also electro-optic in that the application of an external electric field aligns the lithium ion

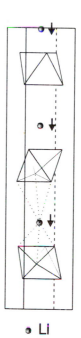

● Li

Fig. 4.11 The LiNbO$_3$ structure showing concerted lithium ion displacements of the ferroelectric phase. Polyhedra outlines represent NbO_6 octahedra.

displacements and also changes the refractive index of the material. This behaviour has led to important applications in optical devices.

4.6   The spinel structure

The spinel structure is widely adopted by ternary oxides and sulphides of the stoichiometry AB_2X_4, X = O,S. The overall charge on the cations is eight and the possibilities for A and B are summarised in Table 4.3. Note that the label B applies to the cation type which is twice as abundant in the structure. The two cations are of a similar size in contrast to the perovskites.

Table 4.3 Possible charges on A and B in spinels.

Cation A charge	Cation B charge	Example
2+	3+	$MgAl_2O_4$
4+	2+	$TiMg_2O_4$, $ZrCu_2S_4$
6+	1+	Na_2WO_4[†]

† Distorted

The spinel structure is rather more complicated than those discussed earlier though one helpful representation is shown in Fig. 4.12. The structure is based upon a cubic close packed arrangement of anions in which a proportion of tetrahedral and octahedral sites is occupied. With *n* close packed anions, 2*n* tetrahedral holes and *n* octahedral holes are generated. In the spinel structure one–eighth of the tetrahedral holes and one–half of the octahedral

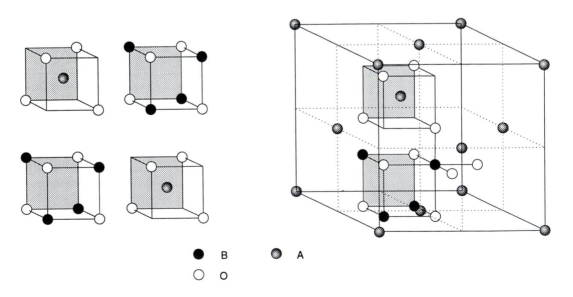

● B   ◉ A

○ O

Fig. 4.12 The spinel structure.

holes are occupied in an ordered manner throughout the structure. The A cations normally occupy the tetrahedral sites and the B cations the octahedral ones giving the stoichiometry $A_{2n/8}B_{n/2}X_n$, which is equivalent to AB_2X_4.

The unit cell in Fig. 4.12 shows the tetrahedral and octahedral co-ordination of the A and B cations. Rather than considering the structure as the filling of holes in a close packed anion array one alternative is to draw the structure as a face centred cubic array of A ions into which smaller cubes of stoichiometry B_4X_4 and AX_4 are added. The AX_4 and B_4X_4 cubelets alternate in the structure and the unit cell contains four of each giving the unit cell stoichiometry $4(AX_4) + 4(B_4X_4) + 4A$ (from the cell corners and faces) $= A_8B_{16}X_{32}$ corresponding to the formula AB_2X_4. The structure is face centred cubic; consideration of any ion in the unit cell will show other ions in identical environments with translations of $+(\frac{1}{2},\frac{1}{2},0)$, $+(\frac{1}{2},0,\frac{1}{2})$ and $(0,\frac{1}{2},\frac{1}{2})$. The powder X-ray diffraction pattern of a spinel, for example Fig. 4.13, shows the typical general absences for a material crystallising with a face centred cubic structure.

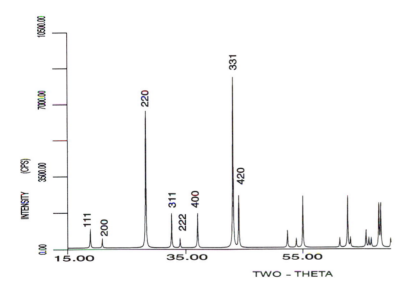

Fig. 4.13 The powder X-ray diffraction pattern of a face centred cubic material such as a spinel. Owing to the high symmetry of the spinel lattice some of the reflections, e.g. 200, are normally absent.

This arrangement of the A and B cations over the tetrahedral and octahedral sites in the spinel structure, that is $A^tB_2^oX_4$, where the superscripts refer to the coordination geometry, is known as the *normal* arrangement and the structure as the *normal* spinel structure. The B ion, which exists at twice the level of the A cation, occupies only octahedral sites within the structure. In some compounds adopting the spinel structure, known as *inverse* spinels, a different arrangement of cations exists corresponding to $B^t(AB)^oX_4$, where all the A cations occupy octahedral sites and the B cations are distributed equally over the two coordination geometries.

Why some materials adopt the inverse spinel structure rather than the normal one is a result of a number of factors including the ion charges. The following discussion applies to the most common formulation $(A^{2+}B_2^{3+})X_4^{2-}$ but the arguments can be equally well extended to the $(A^{4+}B_2^{2+})X_4^{2-}$ system.

Ionic size is one parameter, though flexibility in the position of the anions in the spinel structure permits a range of cation sizes to be coordinated at their preferred contact distances. The higher charged B cations are generally smaller than the A ions, whilst the tetrahedral sites in a close packed anion array are significantly smaller than the octahedral holes. This would point towards the smaller B cations preferring to occupy the tetrahedral sites i.e. the *inverse* formulation $B^t(AB)^oX_4$. However, the oxide ions are not fixed and positional displacements are possible, to some extent, to accommodate differently sized ions.

The ion size factor is generally outweighed by lattice energy calculations; probably the most important parameter in determining the cation distribution. Calculation of the lattice energy of spinels with the same cations, but with the two different distributions, shows that the normal spinel often has a slightly higher lattice energy than the inverse description. The simple reason for this is that lattice energies are determined through summation of all the Coulombic interactions of the ion pairs in the lattice. That is,

$$U_L \propto \Sigma\, z_+ z_- / r_{+-} \tag{4.5}$$

where U_L is the lattice energy, z_+ and z_- cation and anion charge, r_{+-} their separation and the summation is carried out over the whole lattice for all ion pairs. Placing a highly charged cation on a high coordination site contributes strongly to the lattice energy and produces a greater lattice energy than other arrangements. This lattice energy consideration indicates that the majority of (2+,3+) spinels should adopt the normal structure. In practice for non-transition metal systems, where the third factor, discussed below, is non-operative, inorganic spinels, for example $MgTi_2O_4$, are generally found to adopt the normal spinel structure.

Fig. 4.14 d-Orbital distributions in octahedral and tetrahedral fields. Filling of orbitals is for a d^3 ion.

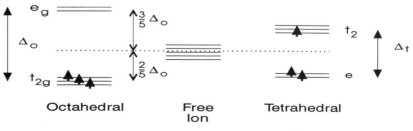

Octahedral   Free Ion   Tetrahedral

a   b

In spinels which contain a first row transition metal, a third factor, partial occupancy of the d-orbitals, is of importance in determining the cation distribution. This is the crystal field stabilisation energy of the various ions in the octahedral and tetrahedral coordination geometries. The occupancy of the d-orbitals on Cr^{3+} ions in a spinel on both tetrahedral and octahedral sites is shown in Figs. 4.14a and 4.14b. In an octahedral site the d-orbitals are split into two groups, the t_{2g} set (d_{xy}, d_{yz} and d_{xz}), which point between the ligands, and the e_g set ($d_{x^2-y^2}$, d_{z^2}), which point towards the ligands. The t_{2g} set are stabilised relative to the e_g set by an amount Δ_o. In the tetrahedral case the arrangement is inverted and the separation of the t_2 and e sets is Δ_t where $\Delta_t \approx 4/9\Delta_o$. In oxide spinels the splitting of the t_{2g} and e_g levels is fairly small as oxide is a weak field ligand. This leads to oxide spinels being high spin materials where the electrons are distributed in the d-orbitals with the maximum number of parallel spins.

With Cr^{3+}, a d^3 ion, on an octahedral site only the t_{2g} set of orbitals are occupied, Fig. 4.14a; owing to the lower energy of these orbitals this octahedral configuration is *stabilised* relative to the case where the ion is non-coordinated to a level known as the crystal field stabilisation energy (CFSE). In terms of Δ_o this CFSE is $3\times2/5\times\Delta_o = 6/5\times\Delta_o$. This may be compared to the case where chromium occupies a tetrahedral site, Fig. 4.14b; here with a high spin electron distribution, $e^2t_2^1$, the CFSE is $2\times3/5\Delta_t - 2/5\Delta_t = 4/5\times\Delta_t$ ($\approx \frac{1}{3}\Delta_o$). This is much less than the octahedral case and the CFSE considerations imply that Cr^{3+} would very much prefer to occupy an octahedral site than a tetrahedral one in a spinel. In a spinel, such as $MgCr_2O_4$, the CFSE considerations reinforce the lattice energy support for the normal spinel with the cation distribution $Mg^tCr_2^oO_4$.

The difference between the CFSE in octahedral and tetrahedral geometries is known as the excess octahedral stabilisation energy and the values for various first row transition metal 2+ and 3+ ions are given in Table 4.4.

Table 4.4. Excess octahedral stabilisation energies (kJ mol^{-1})

Mn^{2+}	Fe^{2+}	Co^{2+}	Ni^{2+}	Cu^{2+}	Ti^{3+}	V^{3+}	Cr^{3+}	Mn^{3+}	Fe^{3+}
0	17	31	86	64	29	54	158	95	0

The high excess octahedral CFSE for Cr^{3+} of 158 kJ mol^{-1} (corresponding to $6/5\times\Delta_o - 4/5\times\Delta_t$) was demonstrated above and supports these 3+ ions on an octahedral site. If another spinel, Fe_3O_4, is considered, the use of CFSE's to stabilise an inverse spinel can be demonstrated. Fe_3O_4 can be rewritten as $Fe^{2+}Fe^{3+}_2O_4$, fitting in with the general spinel formula. The excess octahedral CFSE for Fe^{3+} is zero; as a d^5 ion in a high spin environment the distribution of the d-orbitals, which each contain one electron, produces no CFSE in any coordination. However, for Fe^{2+}, a d^6 ion, there is a small excess octahedral CFSE of 17 kJ mol^{-1}; this indicates that this ion would rather occupy an octahedral site in preference to the tetrahedral site that this 2+ ion would be allocated in a normal spinel structure. In Fe_3O_4 the excess octahedral CFSE

In most spinels other than Fe_3O_4 the excess CFSE would need to be somewhat larger than this to overcome the lattice energy preference for a normal arrangement.

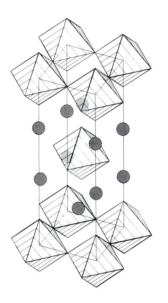

Fig. 4.15 K_2NiF_4 structure shown with NiF_6 octahedra and potassium ions as circles.

for Fe^{2+} is sufficient to overcome the lattice energy mandate and the Fe^{2+} ions occupy octahedral sites giving the inverse spinel $(Fe^{3+})^t(Fe^{2+},Fe^{3+})^oO_4$. With Fe^{3+} and Fe^{2+} occupying identical sites in the lattice, electrons may readily hop from one iron site to the next and Fe_3O_4 is a good electronic conductor.

The zero excess octahedral CFSE for Fe^{3+} means that many transition metal spinels containing this species are inverse, as most 2+ transition metals have some preference for octahedral sites. Hence, $CoFe_2O_4$ and $NiFe_2O_4$ are also inverse spinels. Other inverse spinels in 2+, 3+ systems are rather rare as there must be a significant CFSE driving force to overcome the lattice energy preference for normal spinels. One ion which does frequently form inverse spinels is Ni^{2+} which has a high excess octahedral CFSE of 86 kJ mol^{-1}. In $NiAl_2O_4$ this demand of Ni^{2+} for an octahedral site produces the ion arrangement $Al^t(NiAl)^oO_4$, an inverse spinel.

4.7 K_2NiF_4

The K_2NiF_4 structure is shown in Fig. 4.15 and consists of sheets of NiF_6 octahedra sharing four vertices. These layers are separated by K^+ ions in nine fold coordination to fluorine. The structure is body centred with the NiF_6 octahedron at the unit cell centre displaced by (½,½,½) from that at the cell origin. This structure is widely adopted by ternary oxides of the stoichiometry A_2BX_4 where one cation, A, is much larger than the other, B. This contrasts with the spinel structure which has the same stoichiometry but occurs where A and B have similar ionic radii. Examples include Sr_2TiO_4, La_2NiO_4, Cs_2UO_4 and the superconducting phase $(La_{1.85}Ba_{0.15})CuO_4$ discussed in Chapter 8.

4.8   **Problems**

4.1 Calculate a tolerance ratio for perovskite $PbTiO_3$ taking the ionic radii as 1.68 Å (Pb^{2+}), 0.745 Å (Ti^{4+}) and 1.26 Å (O^{2-}). This material is a ferroelectric. Explain this behaviour in terms of likely ion displacements.

4.2 By consideration of CFSE factors predict whether the following spinels would be expected to adopt the normal or inverse structures. (a) $FeCr_2O_4$ and (b) $NiGa_2O_4$.

5 Electronic, magnetic and optical properties of inorganic materials

The electronic properties exhibited by solids are crucial in a large number of inorganic materials applications. These unique electronic properties result from the extended structures adopted by many inorganic materials, where strong interactions between the atoms, ions or molecules occur throughout the lattice. In terms of conductivity, behaviour ranges from insulating through semiconducting to metallic and superconducting. Similarly, the interaction between the unpaired electrons in inorganic material structures gives rise to magnetic properties characteristic of the whole solid rather than the individual atoms. Finally, the potential for concerted ion displacements in neighbouring unit cells produces ferroelectrics and optoelectric materials. This chapter provides an introduction to these co-operative phenomena.

5.1 Introduction to band theory

In simple molecules the interactions between the atomic orbitals on the various atoms forming the compound give rise to molecular orbitals. The number of molecular orbitals generated in a molecule is identical to the number of atomic orbitals used in forming them. This may be illustrated by consideration of the π orbitals formed in a series of long chain alternate alkenes, Fig. 5.1.

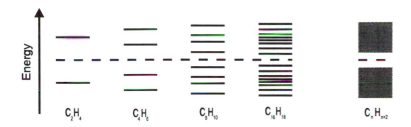

Fig. 5.1 π Molecular orbitals of alternate alkenes. The dotted line separates the bonding and antibonding orbitals.

    In ethene, Fig 5.2a, the p_z orbitals on the two carbon atoms interact to form one bonding and one antibonding configuration. If the number of carbon atoms and π orbitals in the chain is increased to four, producing 1,3 butadiene, Fig. 5.2b, the four carbon π orbitals can interact to give rise to two bonding and two antibonding orbitals. Further extension to an eight

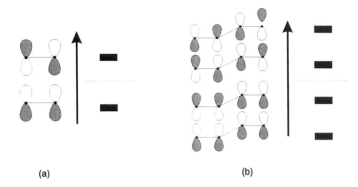

Fig. 5.2 Combinations of p orbitals giving rise to π molecular orbital in ethene (a) and 1,3 butadiene (b).

(a)   (b)

carbon atom chain produces four bonding and four antibonding molecular orbitals. With continued extension of the chain, every pair of carbon p_z orbitals added to the molecular orbital diagram provides one bonding and one antibonding combination. The molecular orbitals, within the bonding and antibonding groups, become closer and closer in energy, and this continues until such time as the chain has an almost infinite number of orbitals, when they form a practical continuum. In such cases, rather than drawing all these energy levels as separate lines, they are replaced in the energy level diagram by a *block* representing a continuous *band* of energy levels. Hence, in the case of an infinitely long chain of π orbitals, corresponding to $(CH)_x$, the energy level diagram consists of two bands, one bonding in character and one antibonding in character.

This behaviour of atomic orbitals in infinite structures, forming continuous bands of energy levels, is frequently found in inorganic materials where there is overlap between the atomic orbitals. The simplest case is that of an elemental metal such as sodium. The electronic structure of a sodium atom is $1s^2 2s^2 2p^6 3s^1$; the core levels with principal quantum numbers one and two are held close to the nucleus and extend poorly into the body centred cubic lattice adopted by sodium. However, the 3s orbital extends out from the nucleus and can overlap with similar orbitals on neighbouring sodium atoms. In crystalline sodium metal, with an almost infinite lattice and the number of orbitals approaching Avogadro's Number 6×10^{23}, the overlap of these 3s orbitals gives rise to a band. Other atomic orbitals on sodium which are sufficiently diffuse, and overlap with similar orbitals on neighbouring atoms, will also give rise to bands, for example, the 3p levels. The band structure of sodium metal derived from the 3s and 3p levels is shown in Fig. 5.3. The width of the individual bands derived from the 3s and 3p orbitals is such that they also overlap.

Remembering that each atomic orbital gives rise to one molecular orbital and, hence, one level in the continuous band, the four orbitals on each sodium atom ($3s$, $3p_x$, $3p_y$ and $3p_z$) will, within the final band structure, give rise to four levels. Each level can contain two electrons as with the original atomic orbitals on the sodium atom. In an infinite lattice of N sodium atoms $4N$ energy levels are generated. Each sodium ion contributes only one electron to this band structure, shown in Fig. 5.3, which can hold eight and, therefore, the band is only partially filled. It is this partially filled band which gives rise to the electronic properties of metals such as high electrical conductivity and lustre. Any inorganic material which has an arrangement of atoms, ions or molecules that produces *partially filled bands* would be expected to demonstrate the physical properties of a metal.

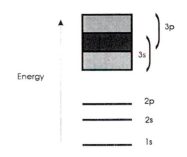

Fig. 5.3 The band structure of sodium metal.

5.2   Band structure and electronic properties

Metals

In all solids with effectively infinite structures there will be an interaction between atomic orbitals with the potential to give a band structure. The electronic properties of a solid depend upon the nature of these bands in terms of their width, separation, the number of electrons they can hold, and the number of electrons in the system. The width of the bands depend on the degree of overlap of the orbitals which interact to give rise to them. Strongly overlapping atomic orbitals give rise to wide bands whilst weakly interacting orbitals give rise to narrow bands. This can be illustrated by reference back to molecular orbitals. Weakly overlapping atomic orbitals give rise to bonding and antibonding molecular orbitals with energies close to those of the original atomic levels, that is with little dispersion in energy terms; with strongly overlapping atomic orbitals the bonding and antibonding combinations are more spread out along the energy axis. The electrons in the wide band can be considered to be dispersed over the whole lattice whilst the electrons occupying a narrow band are more strongly associated with the nuclei.

Band structure diagrams are frequently drawn to represent the number of electrons which may occupy a band. In the band structure derived from the $(CH)_x$ chain there will be a large number of energy levels near the centre of the band and much fewer near the edges. Evidence for this build up of energy levels at the mid point of the energy scale can be seen in the molecular orbital diagram of $C_{16}H_{18}$. The information on the number of energy levels at any particular point in a band can be represented in a density of states diagram such as Fig. 5.4 where $N(E)$, the density of states (or energy levels) is plotted as a function of energy. Filling of the bands by electrons in such a diagram can be denoted by shading of the band. The level to which a band structure is filled is denoted as the *Fermi level*, E_F. If this Fermi levels falls in the middle of a band then a material can be expected to exhibit metallic properties. The band structure of magnesium metal is shown in Fig. 5.5.

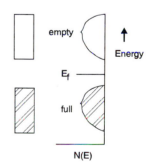

Fig. 5.4 Density of states diagram for $(CH)_x$.

Insulators

If the Fermi level lies at the top of a band it is normally drawn mid way between this band and the next, which will be empty, Fig. 5.6. This band structure is that of an insulator which crystallises with an infinite lattice; diamond is a good example. The separation of the filled and empty bands, the *band gap*, is large, typically greater than 3.0 eV.

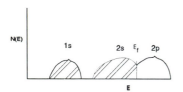

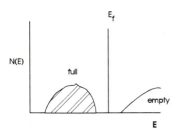

Fig. 5.5 Band structure of magnesium metal as a density of states diagram.

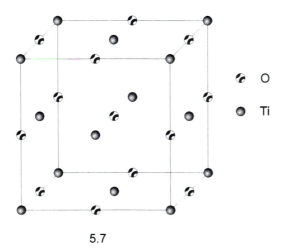

Fig. 5.6 Density of states and Fermi level for an insulator.

Semiconductors

With semiconductors, a similar band diagram to that of insulators is obtained; however in these materials the band separation is much smaller, typically less than 3.0 eV. Silicon crystallises with the same structure as diamond and the band structure is very similar. Differences arise because the atomic orbitals on silicon are higher in energy than those on carbon and also more diffuse; this modifies the band structure to produce wider bands and a smaller separation between the filled (valence) band and the empty (conduction) band. In silicon and similar materials with a small separation of the filled and empty band there is sufficient thermal energy at room temperature to promote some electrons from the valence band to the conduction band. These electrons, and the holes in the valence band that they leave behind, give rise to the moderate conduction properties of semiconductors. The smaller the band gap and the higher the temperature then the greater the number of electrons which can be excited into the conduction bands. This results in higher conductivity in materials with small band gaps and the increasing conductivity seen in semiconductors with increasing temperature.

5.3   Simple applications of band theory to inorganic materials

First row transition metal monoxides

The colours of the majority of transition metal monoxides are quite different from those normally found for the same metal ion in octahedral coordination

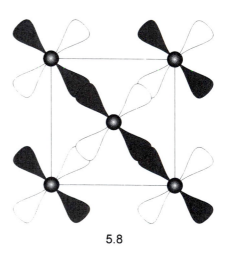

5.7

5.8

Figs. 5.7 and 5.8 Rocksalt structure and the overlap of the metal t_{2g} orbitals in the unit cell faces.

complexes. This is particularly so for the early transition elements, for example, $V(H_2O)_6^{2+}$ is violet whilst VO is black, but for nickel the oxide is the same colour, green, as $Ni(H_2O)_6^{2+}$. For the early transition metals this would indicate that the electronic structure of the oxides is quite different from discrete octahedrally coordinated 2+ ions.

The first row transition metal oxides from TiO to NiO adopt the rocksalt structure shown in Fig. 5.7, though many of these compounds are non-stoichiometric as discussed in Chapter 6. The coordination geometry of the metal ion is octahedral and the d-orbitals on the metal ion will be split energetically into two sets as in typical coordination complexes. The e_g levels, corresponding to the $d_{x^2-y^2}$ and d_{z^2} orbitals, point towards the oxide ions in the structure and the t_{2g} levels, formed from the d_{xy}, d_{xz} and d_{yz} orbitals on the metal, point between the oxide ions. Looking at these directional properties of the t_{2g} levels in relation to the structure of the monoxides, Fig. 5.8, shows that these orbitals point towards identical orbitals on neighbouring metal atoms.

In TiO the Ti^{2+} d-orbitals are very diffuse as the effective nuclear charge felt by these levels is low. This results in good overlap between the t_{2g} orbitals along the (110) directions in the crystal. Hence, rather than considering discrete t_{2g} levels in solid TiO, the electronic structure should be described in terms of a band. The bands formed from the overlap of the t_{2g} orbitals will be able to hold up to six electrons from each contributing titanium ion; Ti^{2+} is a d^2 ion and these two d-electrons will *partially* occupy the t_{2g} band, Fig. 5.9a. TiO, therefore, exhibits metallic behaviour with a high electronic conductivity of about 10^3 $(\Omega\ cm)^{-1}$. Similar electronic behaviour occurs with other early transition metal monoxides, for example VO.

Typical ranges for conductivities are metals 10^1-$10^5(\Omega\ cm)^{-1}$, semiconductors 10^{-5}-10^1 $(\Omega\ cm)^{-1}$ and insulators $<10^{-12}(\Omega\ cm)^{-1}$

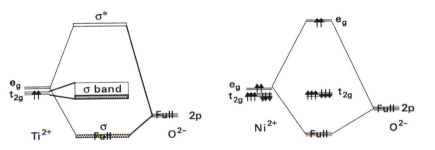

Fig. 5.9 The band structure of TiO (a) and the corresponding levels in NiO (b).

The behaviour of the latter transition metals is quite different. As the first row is crossed, increasingly poor screening of the d-electrons by each other results in a rapidly increasing effective nuclear charge. This causes a marked contraction of the d-orbitals and consequently the overlap between the t_{2g} levels decreases significantly. Rather than forming a band, the t_{2g} levels on the nickel ions within NiO should be considered to be discrete levels. In addition, the increased number of d-electrons for nickel, Ni^{2+} d^8, means that the t_{2g}

levels are filled and it is the e_g metal orbital set which is partially empty, Fig. 5.9b. These e_g orbitals point towards the oxide ions and cannot overlap with those on neighbouring nickel ions to form a band structure. NiO therefore shows no metallic properties. It is an insulator and the d-orbital energies are very similar to those in the hexa-aquo ion, leading to similar d–d transitions and, hence, visible spectrum.

Trioxides

The contrasting electrical properties of WO_3 and ReO_3 can also be rationalised in terms of their band structures. WO_3 is a pale yellow, insulating solid whilst ReO_3 is a beautiful, lustrous, red metal. Both materials crystallise with the ReO_3 structure described in Section 4.1 though that of WO_3 is slightly distorted by twisting of the WO_3 octahedra. The octahedral co-ordination of the metal produces the same d-orbital splitting observed in the monoxides, producing the t_{2g} and e_g sets. The p-orbitals on the oxide ions are also divided into two sets; one p-orbital points towards the metal ion whilst the other two, p_x and p_y, are perpendicular to the M–O–M direction. The t_{2g} orbitals on the metal ions and the p_x, p_y orbitals on the oxide ion overlap as shown in Fig. 5.10 producing a band which can contain up to six electrons. In WO_3, W^{6+} d^0, this band is empty but in ReO_3, Re^{6+} d^1, the one d-electron partially fills the band giving rise to the metallic characteristics of this material. At room temperature, ReO_3 is as good a conductor of electricity as copper metal.

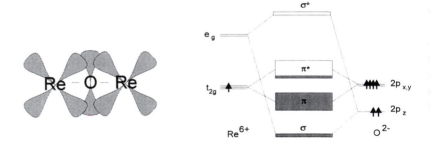

Fig. 5.10 Re (t_{2g}) and O (p) orbital π overlap and the band structure of ReO_3.

The insertion compounds of WO_3, for example Na_xWO_3 or in electronic terms $Na_x^+[WO_3^{(6-x)+}]$, contain the electron donated by the inserted sodium atom in the π band discussed above. These materials, therefore, demonstrate metallic characteristics and $Na_{0.6}WO_3$ is a gold coloured solid.

Graphite intercalates

The band structure of pure graphite is shown in Fig. 5.11a. A particular feature of this diagram is that the valence and conduction bands touch but there are no energy levels at the Fermi level. Another way of considering graphite is as a zero band gap semiconductor. Due to thermal energy a few electrons are excited from the lower valence band into the conduction band and the Fermi level in graphite, above absolute zero, lies slightly into the conduction band. As a result graphite is a moderate conductor of electricity. However, removing or adding electrons to the graphite band structure will produce a Fermi level well inside either the conduction or valence band and this can be achieved chemically through the insertion chemistry of graphite.

Reaction of graphite with potassium to yield C_8K transfers electrons from potassium metal on to the graphite layers; these occupy the valence band and the density of states diagram in Fig. 5.11b results. Similarly reaction with bromine to form C_8Br transfers electrons from the graphite layers on to bromine to form bromide and the valence band is partially emptied, Fig. 5.11c. Both C_8K and C_8Br are true metals and exhibit higher electrical conductivity than pure graphite

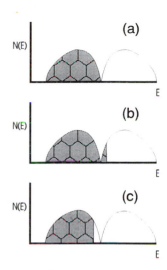

Fig. 5.11 Band structures of (a) graphite, (b) C_8K and (c) C_8Br.

5.4 Magnetism in extended structures

As with the electronic properties of solids the important magnetic properties of inorganic materials result from the interaction between the atomic centres in the extended lattice.

Ferromagnetism, ferrimagnetism and antiferromagnetism

Transition metals in the majority of their compounds have partially filled d-orbitals and frequently a proportion of the electrons occupying these orbitals are unpaired; this gives rise to the magnetic behaviour. Where discrete metal centres exist, for example solutions of transition metal compounds and in many simple, molecular compounds, the unpaired electrons are normally orientated randomly on different metal centres. This random orientation of electron spins and their associated magnetic fields is known as *paramagnetic* behaviour.

However, in solid materials with ordered lattices, the unpaired electrons on individual metal centres can interact, resulting in particular alignments. Paramagnetic behaviour, Fig. 5.12a, is observed in the majority of solid inorganic materials at high temperatures where any interaction of the electrons between metal centres is removed by thermal disordering.

The ordering of the magnetic moments in solids may occur in a variety of ways. If the unpaired electrons align such that their magnetic moments all orientate in a parallel manner the material is *ferromagnetic*, Fig. 5.12b. One alternative is that the magnetic moments may align on neighbouring metal centres in opposite directions or antiparallel, Fig. 5.12c, with the overall magnetic moment averaging to zero: such behaviour is known as *antiferromagnetism*. A third type of behaviour, *ferrimagnetism*, is where the alignment of magnetic moments produces an overall magnetic moment, but not all the individual electron spins are parallel; Fig. 5.12d shows one example.

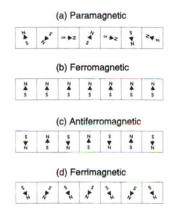

Fig. 5.12 Spin arrangements in paramagnetic, ferromagnetic, antiferromagnetic and ferrimagnetic materials.

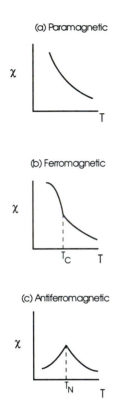

(a) Paramagnetic

(b) Ferromagnetic

(c) Antiferromagnetic

Fig. 5.13 Temperature dependence of the magnetic susceptibility of paramagnetic (a), ferromagnetic (b) and antiferromagnetic materials (c).

The magnetic interaction between the various centres in a solid is normally relatively weak and is readily overcome by thermal energies. Hence, many materials which have magnetically ordered structures at low temperatures become paramagnetic on heating.

The temperature dependence of the magnetic susceptibility of paramagnetic, ferromagnetic and antiferromagnetic materials is shown in Fig. 5.13. The magnetic susceptibility is a measure of the magnetisation of a material when placed in a magnetic field. The positioning of a paramagnetic material with randomly-orientated, individual, magnetic moments in an external magnetic field results in their partial alignment parallel to the field; this acts against thermal effects which tend to disorder them. High alignment of the individual magnetic moments in a sample parallel to the external field produces a high magnetic susceptibility. Hence, as a paramagnetic material is cooled the energy available for thermal disordering of the individual magnetic moments is reduced and such nuclei align more readily parallel to the external magnetic field; Fig. 5.13a results with an increasing susceptibility with decreasing temperature.

If the material undergoes a transition to a state where *all* the individual magnetic moments align, that is a ferromagnetic material, there will be a surge in the magnetic susceptibility as all the electrons attempt to line up in the same direction. The temperature of this transition is known as the Curie temperature, T_C; the temperature dependence of the magnetic susceptibility of a material which becomes ferromagnetic on cooling is shown in Fig. 5.13b. Note that above the Curie temperature the compound acts exactly as a paramagnetic material.

If a material undergoes a transition to an antiferromagnetic state the susceptibility drops markedly, Fig. 5.13c, as the individual magnetic moments try to line up half parallel and half antiparallel to the external field. The temperature at which a material becomes antiferromagnetic is known as the Néel temperature, T_N.

5.5   Magnetic properties of oxides

Superexchange

One mechanism which causes magnetic ordering in inorganic materials is superexchange and is illustrated in Fig. 5.14. Many inorganic systems, for example oxides and sulphides, contain metal centres separated by an anion. A good example is MnO which adopts the rocksalt structure where overlap of the metal e_g d-orbitals with the p-orbitals on the anion can occur along each of the cell axes. Mn^{2+} has five d-electrons in a high spin configuration and two of these occupy the e_g levels. An unpaired electron occupying an e_g orbital on Mn^{2+} can interact with an electron in the oxide p-orbital forming a pair of antiparallel spins. The other electron in the oxygen p-orbital is necessarily of opposite spin orientation and this can in turn interact with an e_g electron on the next manganese ion. This chain of interactions can run throughout the structure and will produce alternating spin orientations on all

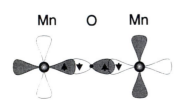

Fig. 5.14 Schematic of superexchange via an oxide ion.

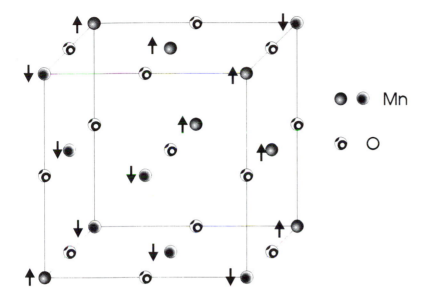

Mn

O

Fig. 5.15 Antiferromagnetic ordering of Mn^{2+} ions in MnO below 122 K.

neighbouring manganese ions. At low temperatures the strength of these superexchange interactions through oxide ions is such that this magnetic ordering process occurs and MnO is antiferromagnetic below 122 K, T_N, Fig. 5.15. Above the Néel temperature the thermal energy is greater than the interaction energy between the unpaired electrons and the electrons on adjacent manganese ions become randomly oriented.

Similar behaviour to that of MnO is seen with other late first row transition metal monoxides; $Fe_{1-x}O$, CoO and NiO, all adopting the rocksalt structure, with T_N's of 198, 293 and 523 K respectively. The increasing Néel temperature across this series may reflect the decreasing lattice parameters and increasing overlap between the metal and oxide ion orbitals.

Spinels

The ferrite spinels of composition MFe_2O_4, M = Fe, Ni, Zn, have important magnetic properties which have led to their use in transformer cores and in magnetic recording media. The distribution of cations in the spinel structure has been discussed in Section 4.6 and spinels containing Fe^{3+} because of its zero excess octahedral CFSE are often inverse. This results in structures which contain Fe^{3+} on both tetrahedral and octahedral sites. In Fe_3O_4 the iron ion distribution is that of an inverse spinel, $(Fe^{3+})^t(Fe^{2+},Fe^{3+})^oO_4$, and all the metal ions have unpaired electrons. The superexchange interactions between the various sites is complex but the overall effect, in the magnetically ordered phase, is to align the magnetic moments of all the octahedral sites in one direction and those on tetrahedral sites in the opposite. Fe_3O_4 is, therefore, a ferrimagnet.

5.6 Optical properties of solids

Non-linear optical materials

The interaction of laser light with the electronic charge distribution in non-centrosymmetric crystals gives rise to observable non-linear effects which may be used for the amplification, modulation and conversion of the laser frequencies. Frequency conversion is particularly useful as it extends the range of available laser frequencies. In designing inorganic materials as non-linear optic materials the relationship between the charge distribution, dictated by the structure, and the optical properties is paramount.

The interaction of light with the electronic charge distribution around an ion induces oscillations of the electron cloud. With light of relatively low intensity the induced polarisation, the displacement of the electron cloud or dipole moment per unit volume, P, is directly proportional to the magnitude of the electric field of the light wave, E. That is

$$P = \chi E \qquad (5.1)$$

where χ is the linear optical susceptibility, a function of the refractive index of the material. In materials with a crystal symmetry below tetragonal three values of χ are required to define the optical susceptibility in the different lattice directions. Laser light generates very intense electric fields and gives rise to non-linear optical effects and the expression for P must be modified to

$$P = \chi^{(1)} E + \chi^{(2)} E^2 + \chi^{(3)} E^3 + \dots \qquad (5.2)$$

where $\chi^{(2)}$ and $\chi^{(3)}$ are constants, the second and third order susceptibilities respectively. The second term, including $\chi^{(2)}$, is responsible for the generation of second harmonic radiation with a wavelength half that of the impinging laser light. A large $\chi^{(2)}$ represents a material which will produce a higher level of this frequency doubled radiation; in order to compare different materials, values of $\chi^{(2)}$ are often quoted relative to a standard such as potassium dihydrogen phosphate, KH_2PO_4 with $\chi^{(2)}=1$. As with χ, in low

Fig. 5.16 The structure of $KTiOPO_4$ shown as linked TiO_6 octahedra and PO_4 tetrahedra forming cavities containing potassium ions (circles).

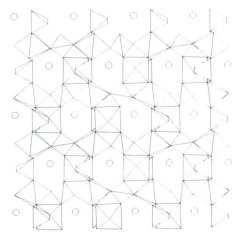

symmetry materials $\chi^{(2)}$ values should be specified for the different crystal directions.

Non-linear optical crystals are generally insulators transparent to the laser light of interest. Non-linear optical effects occur in non-centrosymmetric materials with highly polarisable crystals which also have a high refractive index; a number of non-linear optic crystals develop polarisation as a result of the displacement of ions as in ferroelectrics.

Two important non-linear optic materials are $LiNbO_3$ and $KTiOPO_4$. The structure of lithium niobate was described in Section 4.5. This material has a high refractive index and a high polarisability as a result of the lithium ion displacements in the ferroelectric phase and $\chi^{(2)}$, relative to $KH_2PO_4=1$, is 13. Applications of lithium niobate in optical devices include frequency doublers and waveguides. $KTiOPO_4$, has the structure shown in Fig. 5.16, and is constructed from linked TiO_6 octahedra and PO_4 tetrahedra. The potassium ions occupy cavities within this framework. In the ferroelectric phase the potassium ions are displaced from the centres of these cavities producing a high dipole moment. This material also has a high refractive index. The combination of these properties produces a high value for $\chi^{(2)}$ in eqn 5.2, of 15, relative to $KH_2PO_4=1$, and potassium titanyl phosphate is rapidly developing in non-linear optic applications.

5.7 Problems

5.1 ReO_3 forms the insertion compound $Li_{0.3}ReO_3$. Describe the band structure of this material and the expected electron conduction properties.

5.2 The rutile structure consists of edge sharing MO_6 octahedra as shown in Fig. 5.17. Using this description, which d-orbital on the metal can be used to form a band through overlap with similar d-orbitals on neighbouring metal atoms? Would you expect TiO_2 or VO_2 to exhibit metallic properties?

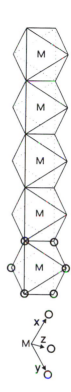

Fig. 5.17 Edge linked MO_6 octahedra along the *c* direction forming the rutile structure. Using the metal centre as the origin the cartesian *x*, *y* and *z* axes point towards the oxide ions as shown.

6 Non-stoichiometry

6.1 Point defects

All crystalline materials contain a certain number of defects, that is their structures are not perfect. The description of unit cells used in Chapter 1 was based around a perfect, ordered array of atoms and molecules. However, real materials do not have structures with a flawless arrangement of atoms and the reason for this can be illustrated by thermodynamic considerations. Consider a crystal into which a number of defects, examples would be a misplaced ion or a vacant site, are gradually introduced; the free energy of the system (ΔG) will fall initially before rising as shown in Fig. 6.1. The reason for this behaviour is the competing effects of the enthalpy (ΔH) and the entropy (ΔS) of introducing the vacancy. Removing ions and molecules from a structure will, typically, reduce the lattice energy; lattice energies are dependent upon the total of the electrostatic attractions between all atom pairs. Hence, the incorporation of defects into a structure will have an enthalpy penalty and ΔH climbs steadily as the number of defects increases. Introducing defects randomly into the structure will produce disorder, so ΔS will be positive. ΔS for each defect is large at first, as the perfect lattice is strongly disrupted, but as more and more defects are introduced further disruption of the lattice is less marked. The overall variation in ΔG, as a function of the number of defects, shows that there will be a certain number of defects corresponding to the minimum in the plot.

For materials such as NaCl and MgO, with high lattice energies and thus a large enthalpy penalty for defect formation, ΔH rises very rapidly and the minimum in ΔG lies at a relatively low number of defects. At room temperature the number of defects in NaCl is about 1 ion in 10^{15}, though as the temperature rises $T\Delta S$ becomes more significant and the number of defects increases rapidly. In other materials the lattice energy costs of introducing defects may be much lower and the equilibrium number of defects consequently rather higher.

Two types of defect are commonly found in ionic inorganic solids and because they do not affect the stoichiometry of the compound they are known as *intrinsic* defects.

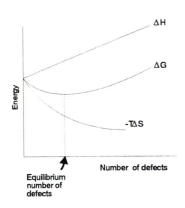

Fig. 6.1 Energetics of defect incorporation.

Schottky defects

Schottky defects, Fig. 6.2, are *vacancy pairs*: ions are removed from their lattice sites and, in order to maintain charge balance, this must be carried out for cation/anion pairs. The pairs of vacancies may associated with each other on neighbouring sites or be quite well separated. This type of defect is commonly found at low concentrations in alkali metal halides.

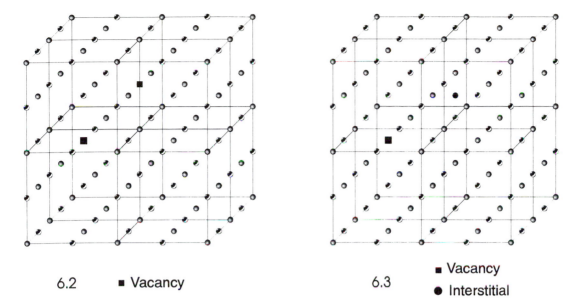

6.2   ■ Vacancy

6.3   ■ Vacancy
        ● Interstitial

Figs. 6.2 and 6.3. Schottky and Frenkel defects in a rocksalt structure.

Frenkel defects

Frenkel defects, Fig. 6.3, are *interstitials*, that is, a displaced atom, ion or molecule that occupies a site within the structure that is not normally filled; the displaced species leaves behind a vacant site. In order for an ion to fill an interstitial site it needs to be small and, because it is quite often close to ions of the same charge, polarisable. Typical materials exhibiting Frenkel defects are the silver halides where Ag^+ forms the interstitial. Again the defects can occur in pairs, that is, a vacancy and interstitial close to each other, or they may be separated by larger distances.

Split interstitials

In the description of Frenkel and Schottky defects no account has been taken of what happens to the lattice around a defect once it has been introduced. In fact the lattice relaxes to accommodate the vacancy and/or interstitial. In the case of an interstitial the ions surrounding the additional ion will move, in so far as it is possible within the constraint of the crystal structure, to fit around it. In some materials the interstitial and an atom displaced from its normal site occur as a pair; this is known as a *split interstitial*, Fig. 6.4.

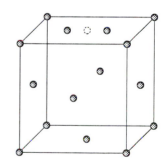

Fig. 6.4 A split interstitial in a face centred cubic metal lattice.

6.2   Non-stoichiometry

In non-stoichiometric materials defects of the type described above occur but to a much larger extent. In addition, the vacancies and interstitials which exist are not formed in pairs. For example, cation vacancies are not balanced by an equal number of anion vacancies as in Schottky defects so, for an overall charge balance, one component of the lattice, normally a metal ion, must be

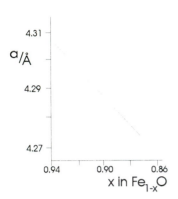

Fig. 6.5 Variation of the lattice parameter of $Fe_{1-x}O$ as a function of x.

oxidised or reduced. Introduction of additional anions on to interstitial sites in UO_2 to form UO_{2+x} with $x=0.1$ requires the uranium oxidation state to increase from 4 to an average of 4.2+. The occurrence of non-stoichiometry is, therefore, restricted to compounds containing a species which is readily oxidised or reduced. Non-stoichiometric behaviour is most frequently found for transition metal compounds though it is also known for some lanthanide, actinide and B metal compounds.

Non-stoichiometric materials are characterised by two criteria. From a thermodynamic standpoint, the free energy of the system depends upon both the composition and temperature. A more useful criterion in that the behaviour is readily identified by experiment, is that the lattice parameter of the system varies smoothly as a function of composition, e.g. Fig. 6.5. Thus, as defects are introduced into a structure, the lattice parameter will change *gradually* between two end members of the non-stoichiometric system. This means that across the full composition range of the non-stoichiometric material the same structure is adopted and that the defects are *randomly* distributed throughout the material. This behaviour should be contrasted with a series of phases of different fixed stoichiometry that may be formed between two elements, e.g. V_2O_3, VO_2 and V_2O_5, all having different structures and, hence, lattice parameters.

The methods by which defects, either vacancies or interstitials, can be introduced in to a general binary compound AB are summarised in Table 6.1. All these types of non-stoichiometric material are known and examples of each type are discussed in the following sections.

Table 6.1 Scheme for the formation of non-stoichiometric compounds from AB

	AB		
Oxidise metal		Reduce metal	
Excess anions AB_{1+x}	Metal vacancies $A_{1-x}B$	Excess metal $A_{1+x}B$	Anion vacancies AB_{1-x}
UO_{2+x}, CeH_{2+x}	$Fe_{1-x}O$	$Zn_{1+x}O$	TiO_{1-x}

Systems in which the metal is oxidised are by far the more common and the level of non-stoichiometry is frequently much larger. The reasons for this are thermodynamic. Oxidation of the metal results in a much smaller enthalpy penalty for the inclusion of defects as, with an increased metal charge and (consequently) smaller metal ions, the lattice energy of the solid increases to partially outweigh the energy required to form the defect. Hence, in Fig 6.1 the ΔH line rises less rapidly than in the case where the metal is reduced and the equilibrium number of defects is, in general, much larger.

6.3   Structures and non-stoichiometry

UO_{2+x}

Stoichiometric UO_2 crystallises with the fluorite structure shown in Fig. 6.6. UO_2 will react with oxygen to form the non-stoichiometric materials UO_{2+x} with $0 < x < 0.25$. $UO_{2.25}$ or U_4O_9 has an ordered (non-random) arrangement of both uranium and oxygen and is thus a stoichiometric phase forming the end member of the series. Between UO_2 and U_4O_9 variable amounts of oxygen are added to the lattice, with a resultant gradual change in the lattice parameter.

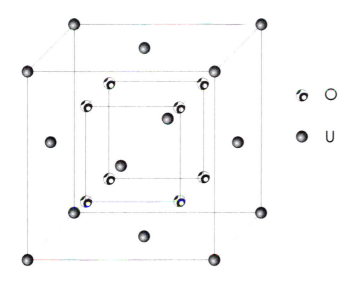

Fig. 6.6 The fluorite structure of UO_2.

Inspection of the UO_2 structure for large, vacant sites which could accommodate an additional oxide ion indicates a possible position at the unit cell centre $(\frac{1}{2},\frac{1}{2},\frac{1}{2})$. The immediate environment of this site shown in Fig. 6.7 consists of eight existing oxide ions at the corners of a cube. The distance from the cell centre to these oxide ions is about 2.00 Å which is somewhat shorter than twice the ionic radius of O^{2-} indicating that even this site is too small to accept, directly, an oxide ion.

Investigations of the structure of UO_2 using powder diffraction techniques have located the oxygen positions in the unit cell of UO_{2+x}. Two observations were made: vacancies existed on the normal oxide sublattice and two types of additional oxide ions were located close to $(\frac{1}{2},\frac{1}{2},\frac{1}{2})$ but displaced in the 110 and 111 lattice directions. Fig. 6.8. shows the two interstitial oxide positions, termed O' and O'' respectively, which, by consideration of the ionic radii, must be associated with vacancies on the adjacent normal oxide ion sublattice. Interstitial oxygen may, therefore, be accommodated in the fluorite lattice but at the expense of oxide vacancies. In order for an overall increase in oxygen stoichiometry these interstitials and vacancies must cluster together. One possibility is shown in Fig. 6.9 with two vacancies being

Note that the fluorite structure is face centred cubic and sites equivalent to that at $(\frac{1}{2},\frac{1}{2},\frac{1}{2})$ occur along the cell edges at $(\frac{1}{2},0,0)$, $(0,\frac{1}{2},0)$ and $(0,0,\frac{1}{2})$.

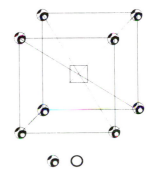

Fig. 6.7 Oxide ion sublattice in the UO_2 unit cell.

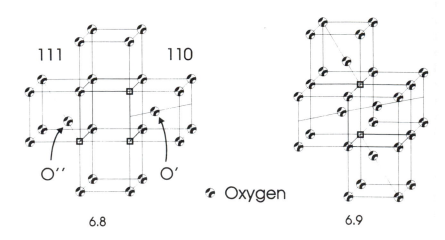

Figs. 6.8 and 6.9 Interstitial oxide ion positions in UO$_2$ and a defect cluster formed from them. The boxes are the unit-cell, oxide-ion sublattice taken from Fig. 6.7.

associated with two O′ type and two O″ type interstitials. The overall composition of this structural block, consisting of eight UO$_2$ units ('U$_8$O$_{16}$') is, therefore, increased by two oxide ions giving U$_4$O$_9$. This agglomeration of interstitials and vacancies is termed a *defect cluster* and represents an energetically favourable way of incorporating large numbers of defects in a non-stoichiometric compound. Other possible clusters are possible; for example extension of Fig. 6.9 by an additional vacancy would permit two more O′ type interstitials to be incorporated into the defect cluster.

The observation that the end member of the non-stoichiometric range is U$_4$O$_9$ results from a structure which contains one additional oxygen in each unit cell. The additional oxide ions are now *regularly* distributed, rather than randomly as in UO$_{2+x}$, 0<x<0.25.

The widespread adoption of the fluorite structure by materials with the AX$_2$ stoichiometry and its open nature leads to frequent observation of non-stoichiometric materials similar to UO$_{2+x}$. Other examples include CeH$_{2+x}$ and materials such as Ca$_{1-x}$Y$_x$F$_{2+x}$ where the fluorite structure of CaF$_2$ has been doped with YF$_3$. Calcium and yttrium share the cation positions in the fluorite structure and the additional fluoride ions occupy similar sites to those of oxygen in UO$_{2+x}$.

Fe$_{1-x}$O

The Fe$_{1-x}$O system is probably the most extensively studied non-stoichiometric system despite the fact that Fe$_{1-x}$O is metastable at room temperature. This means that the phase, formed only at high temperature, requires quenching (cooling quickly) to room temperature in order to leave the iron and oxygen atoms locked into the Fe$_{1-x}$O structure. With increasing temperature the range of non-stoichiometry in this system increases; at 600°C, x can only adopt values to give the range Fe$_{0.93}$O to Fe$_{0.94}$O, at 1000°C the range becomes Fe$_{0.87}$O to Fe$_{0.96}$O.

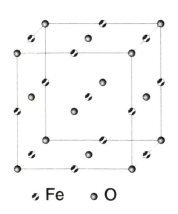

Fig. 6.10 The rocksalt structure of 'FeO'.

In common with most first row transition metal monoxides $Fe_{1-x}O$ adopts the sodium chloride structure shown in Fig. 6.10, though obviously a large number of defects need to be incorporated into the lattice to produce the stoichiometry $Fe_{0.93}O$. The method of incorporating the iron defects is not simply a matter of removing iron from the normally fully occupied sublattice. Powder diffraction measurements on $Fe_{1-x}O$ have shown that some iron atoms occupy interstitial sites within the sodium chloride structure and that the level of iron vacancies on octahedral sites is greater than would be required by the compound stoichiometry alone. The occupied interstitial positions within the lattice were found to be tetrahedral sites such as (¼,¼,¼).

Magnetic measurements have shown that the tetrahedral iron atoms have a charge of 3+. The basic mechanism by which defects are incorporated into '$Fe_{1-x}O$' is illustrated in Fig. 6.11. A number of vacancies are created on the normally occupied octahedral sites but over the whole structure this number is greater than required by the compound stoichiometry. The additional iron required occupies tetrahedral sites adjacent to the vacancies. In order to maintain charge balance the interstitial iron is oxidised to 3+ and a number of iron atoms surrounding the defect but on normal lattice sites must also be oxidised to Fe^{3+}. Note that Fe^{3+}, as a high spin d^5 ion, has no preference for octahedral or tetrahedral sites.

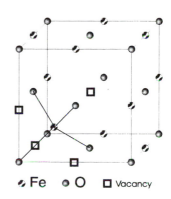

 Fe   O   □ Vacancy

Fig. 6.11 A 4:1 defect cluster in $Fe_{1-x}O$.

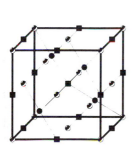

• Interstitial iron
 Oxygen
 Normal iron site
■ Iron Vacancy

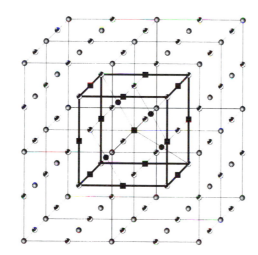

Fig. 6.12 The Koch–Cohen cluster of $Fe_{1-x}O$.

This defect cluster with four vacancies and one interstitial is termed a 4:1 cluster and is believed to occur for low concentrations of iron vacancies, i.e. small x in $Fe_{1-x}O$. At higher defect concentrations a larger cluster consisting of 13 vacancies and 4 interstitials illustrated in Fig. 6.12 has been suggested. This cluster is named after its proposers as a Koch–Cohen cluster. Once again this cluster must be surrounded by a sheath of Fe^{3+} ions occupying normal octahedral sites, to provide charge balance.

$Zn_{1+x}O$

As discussed above, non-stoichiometric systems in which the metal is reduced have a much more limited range of stoichiometry and are rather rare. One example of a reduced metal system is $Zn_{1+x}O$ but x is very small, $\approx10^{-5}$, and moderate values are found only at high temperatures where the formation of defects is favoured through entropy considerations. When zinc oxide is heated the compound loses oxygen from the surface forming $Zn_{1+x}O$; charge balance requires the zinc atoms to be reduced to Zn^+ or Zn^0. These ions migrate from the surface to interstitial positions within the lattice as shown in Fig. 6.13. As the concentration of these interstitial ions is so low, structural studies have been unable to locate their position, but consideration of the wurtzite structure adopted by ZnO and the larger size of these ions, Zn^+/Zn^0, would indicate that they would probably occupy octahedral holes. The formation of lower charged zinc ions/atoms permits an electron transfer process between the various zinc centres. The energy of these transitions occurs near to the visible region of the spectrum giving rise to the yellow colour of hot ZnO. As the number of interstitial zinc ions decreases markedly at low temperatures the colour of ZnO at room temperature is white.

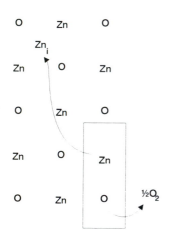

Fig. 6.13 $Zn_{1+x}O$.

TiO_x

Titanium monoxide shows an enormous non-stoichiometry range; at 1400°C the full range of materials from $TiO_{0.7}$ to $TiO_{1.25}$ can be prepared. Over this whole composition range the material adopts structures based on that of sodium chloride. At low oxygen content vacancies exist on the oxide sublattice, whilst in $TiO_{1.25}$ it is the titanium sublattice that has vacancies and

Fig. 6.14 Vacancy distribution in 'TiO'.

this compound can be rewritten as $Ti_{0.8}O$. Even the 1:1 stoichiometry material, TiO, has a structure containing defects with equal numbers of vacancies on both the titanium and oxygen sublattices.

The structure of the limiting composition $TiO_{0.7}$ has not been determined in detail and indeed the structural chemistry of the TiO_x system is very complex. At high temperatures, above 900°C, the vacancies in TiO are randomly distributed and the material is cubic. Below 900°C ordering of the vacancies can occur and in 'TiO' 50% of cations and anions are alternately absent in every third 110 plane, Fig. 6.14.

6.4   Non-stoichiometry in more complex systems

The four non-stoichiometric systems discussed above in terms of their structures, illustrate the major ways in which large compositional variations can be incorporated into materials through defects and interstitials. Non-stoichiometry is a very widespread phenomenon in solid state oxide, sulphide and halide chemistry, but it is also a feature of intercalation chemistry by the nature of the synthesis process.

In more complex oxides, for example ternary phases, and insertion compounds, a detailed level of structural information, as discussed above for a number of binary systems, is often not available and the structures of compounds are often described in terms of an *average* distribution of ions or molecules. To illustrate this a good example is provided by the ternary oxide system $Sr_3Fe_2O_{7-y}$, $0<y<1$.

The structure of $Sr_3Fe_2O_7$ is shown in Fig. 6.15 and can be considered to be blocks of perovskite separated by layers of SrO with the rocksalt structure. The oxygen stoichiometry of this material may be reduced drastically by gradual removal of oxide ions from the site within the perovskite block. Complete removal of oxygen from this site results in the composition $Sr_3Fe_2O_6$ with reduction of the iron, from Fe^{4+} in $Sr_3Fe_2O_7$, to Fe^{3+}. The coordination geometry of the iron decreases from six in an octahedral arrangement to five forming a square based pyramid. For the intermediate compositions in this system, $Sr_3Fe_2O_{7-y}$, $0<y<1$, the structure would be described in terms of partial filling of the oxide ion site between the two iron atoms. Throughout the structure there would be a random distribution of linked FeO_6 octahedra and FeO_5 square pyramids.

Similar changes in the oxygen content of materials, through partial filling of a site, is important in the chemistry of the high temperature superconductors discussed in Chapter 8.

6.5   Elimination of defects, crystallographic shear

In the structural chemistry of UO_{2+x} and $Fe_{1-x}O$ the formation of defect clusters has been observed though these clusters are still *randomly* distributed throughout the structure over the full stoichiometry range. Some materials

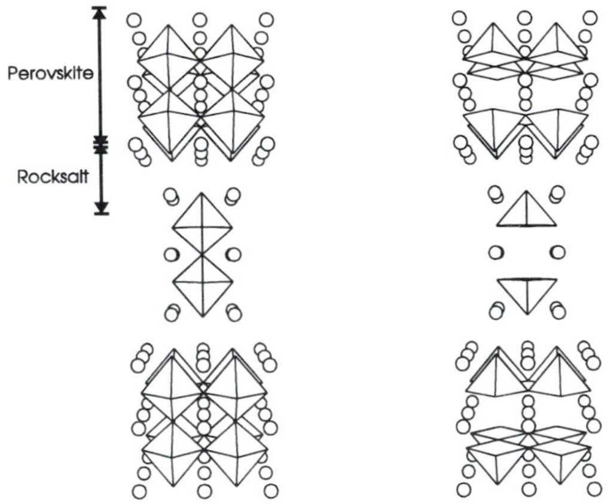

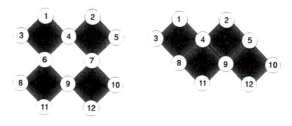

Fig. 6.15 The structures of $Sr_3Fe_2O_7$ and $Sr_3Fe_2O_6$. FeO_6 octahedra and FeO_5 square pyramids are shown as polyhedral outlines and the Sr ions as circles.

assimilate a large number of defects into their structures by incorporating them as an embedded portion of a different structure. This effectively eliminates the defects, forming a new ordered array and generating a more complex structure. This process is known as crystallographic shear and is illustrated schematically in Fig. 6.16.

Crystallographic shear is well characterised in systems derived from corner sharing octahedra and the defects are eliminated by increasing the coordination number of oxygen. The ReO_3 structure, Fig. 4.3, is formed from octahedra sharing all corners. Oxygen deficiency introduced into this structure can be accommodated through the crystallographic shear process as illustrated in Fig. 6.17. Removal of oxygen (labelled 6 and 7) along the line of corner sharing octahedra is followed by a translation along and perpendicular to the line of missing oxygen by the lower set of octahedra.

The coordination number of the oxide ions along the crystallographic shear plane (e.g. the oxide ion labelled 9) is increased to three while that of the metal atom is maintained at six, consistent with the decrease in oxygen level. The method of incorporation of this crystallographic shear into the full ReO_3 lattice is illustrated in Fig. 6.18.

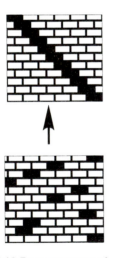

Fig. 6.16 Rearrangement of defects in crystallographic shear.

Fig. 6.17 Oxygen defect elimination in a portion of the ReO_3 structure.

Fig. 6.18 Crystallographic shear along 310 in the ReO₃ structure.The left hand side shows the perfect ReO₃ structure prior to elimination of defects along the 310 plane (diagonal line).The right hand side shows the sheared structure produced with edge sharing octahedra.

The overall stoichiometry of the resultant compound depends upon the frequency of occurrence of the crystallographic shear plane throughout the structure: indeed by changing this frequency the stoichiometry can be varied. If the crystallographic shear shown in Fig. 6.18 is separated by ten normal MO_6 octahedra the compound stoichiometry becomes $M_{10}O_{29}$ and this structure is adopted by $W_{20}O_{58}$. A separation of nine normal MO_o octahedral blocks would result in the stoichiometry M_9O_{26}.

6.6   Intercalation compounds — staging

Whilst many intercalation compounds are truly non-stoichiometric with a random distribution of intercalated species over a proportion of sites, the insertion compounds of some layer systems exhibit a phenomenon known as staging. The potassium intercalates of graphite clearly illustrate this behaviour. In C_8K, the highest potassium intercalate or first stage compound, the potassium ions occur between *all* the carbon layers on sites shown in Fig. 6.19.

The other compounds in the potassium–graphite system have the stoichiometries $C_{24}K$, $C_{36}K$ and $C_{48}K$. In $C_{24}K$ rather than removing a portion of the potassium ions from all the layers, alternate potassium layers are removed completely and one third of the potassium ions from the remaining strata. The potassium ions are disordered within the layer and this compound is known as a second stage intercalate, Fig. 6.19. The $C_{36}K$ and $C_{48}K$ compounds, third and fourth stage intercalates, Fig. 6.19, have potassium ions every third and fourth graphite layer respectively. This behaviour is known as *staging* and is clearly demonstrated in the powder X-ray diffraction patterns. The layer structure of graphite leads to a strong 001 reflection which is readily observed in the powder X-ray diffraction pattern. The position of the 001 reflection for some of the staged compounds is shown in Fig. 6.20 and corresponds to the *c* lattice parameter marked in Fig. 6.19.

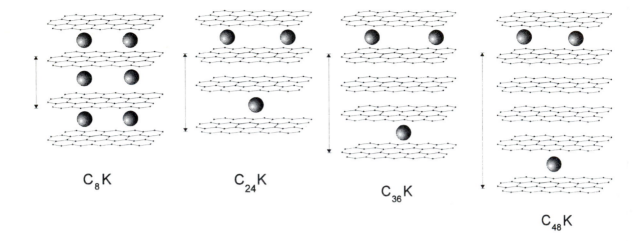

C_8K $C_{24}K$ $C_{36}K$ $C_{48}K$

Fig. 6.19 Staging in C_nK. The *c* lattice parameter is marked for each phase.

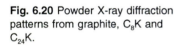

Fig. 6.20 Powder X-ray diffraction patterns from graphite, C_8K and $C_{24}K$.

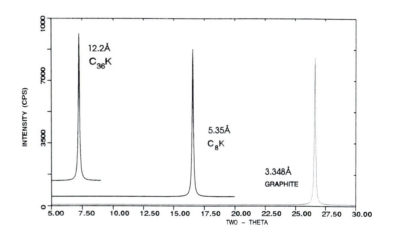

6.7 Problems

6.1 The 111 reflection from CeH_2 (λ=1.54 Å) occurs at 27.96°. Taking the ionic radius of H^- as 2.08 Å calculate the dimensions of the largest sphere which could occupy the (½,½,½) site without removal of hydride ions from their normal sites. The powder X-ray diffraction pattern of $CeH_{2.15}$ is identical to that of CeH_2 except for a shift of peak positions to slightly lower 2θ values. Explain.

6.2 How would you expect the lattice parameters of $Sr_3Fe_2O_{7-y}$ to vary as a function of *y*?

7 Zeolites, intercalation in layer materials and solid electrolytes

The preceding chapters have dealt mainly with the structure and properties of metal oxides, an extremely important group of compounds which illustrate many of the aspects of inorganic solid state chemistry. This chapter provides an introduction to some other widely studied inorganic materials which have significant applications and properties.

7.1 Zeolites

The zeolites are a group of compounds, many of which are naturally occurring minerals, named after their property of evolving water when heated (Greek, *zeo* to boil, *lith* stone). These materials are widely used for their ion exchange, absorption and catalytic properties.

The compounds are characterised by open structures which may incorporate a range of small inorganic and organic species. The frameworks forming the channels and cavities are constructed from linked tetrahedra and many elements, which form TO_4 groups in their structural chemistry, can take the part of the building blocks in zeolites. Examples include AlO_4, SiO_4, PO_4, BeO_4, GaO_4, GeO_4 and ZnO_4. The most common zeolitic materials are based on silicon and aluminium MO_4 tetrahedra linked together. The different ways in which these tetrahedra are connected in three dimensional space give rise to the multitude of different zeolites. At present over 200 aluminosilicate framework structures are known; about 40 of these are naturally occurring minerals.

A general formula for zeolites may be obtained by starting from pure silica which in its various structures, quartz being one example, consists of SiO_4 tetrahedra sharing all vertices. Replacement of some of the SiO_4 tetrahedra by AlO_4 tetrahedra can occur provided that charge balance is maintained by simultaneously incorporating cations into the structure, that is 'SiO_2' is replaced by '$MAlO_2$' or '$M_{0.5}AlO_2$' where M is a mono or divalent cation. The cations introduced alongside the aluminate tetrahedra are frequently hydrated leading to a general formula for zeolites of

$$\{ [M^{n+}]_{x/n} \cdot [mH_2O] \} \{ [AlO_2]_x [SiO_2]_{1-x} \}$$

with the three components, cavity cations, absorbate (water) and framework. The final {} grouping describes the tetrahedral species forming the zeolite framework whilst the first large bracket describes the species present in the cavities. In this description the absorbed species, coordinating to the cation, is water, though as will become apparent, other small molecules are readily

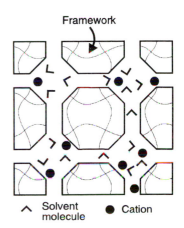

Fig. 7.1 Schematic of the components of a zeolite.

Framework

⋀ Solvent molecule   ● Cation

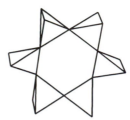
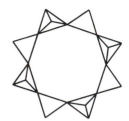

Fig. 7.2 Four, six and eight membered rings formed from linking tetrahedra.

The nomenclature of zeolites is rather unsystematic. Some materials, e.g. sodalite and faujasite, are named after minerals, others e.g. ZSM-5 and VPI-5 were named by researchers during a programme of work to synthesise new frameworks. Thus ZSM was the fifth product of the Zeolite Socony–Mobil project. Zeolites are often described in the form M-[zeolite] where M refers to the cation in the particular [zeolite] framework.

absorbed into the zeolite pores.

Fig. 7.1 can be used to represent schematically a zeolite with the cavities and pores, formed from the tetrahedra, containing the hydrated cations. Following dehydration of a zeolite, readily accomplished by heating under vacuum, the cations migrate to the edges of the cavities and channels, interacting more strongly with the framework oxygen atoms. The cavities in zeolites, particularly once dehydrated, may absorb small molecules other than water. The larger the cavities and pores formed by the framework the larger the molecules which may be absorbed inside the zeolite.

7.2   Zeolite structures

The linking of the TO_4 tetrahedra may occur in numerous ways to produce the huge variety of zeolite structures, but a number of structural features are common to these materials. Linking of the tetrahedral species may take place

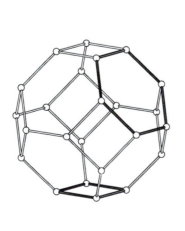
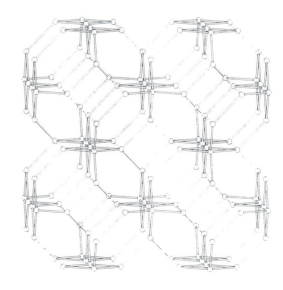

Fig. 7.3 A sodalite or ß cage and the sodalite structure where the ß cages share six-membered rings. Each line connects the silicon/aluminium centres and one six and one four membered ring, corresponding to those in Fig. 7.2, are outlined.

to form four, six and eight membered rings as shown in Fig. 7.2 and these units are then used to form the three dimensional zeolite structures. The wealth of zeolite structures is derived not only from the multitude of ways in which the tetrahedra can be linked, but also from the flexibility of the T–O–T bond angle which can assume values in the extensive range 120–180°.

Figure 7.3 shows a ß or sodalite cage produced from four and six membered rings. This cage frequently occurs itself as a building block in the structures of other zeolites such as zeolite A, Fig. 7.4a, faujasite, Fig. 7.4b and VPI-5, Fig. 7.4c. In this series of materials the sodalite cages are separated by increasing distances. In sodalite the ß cages form a body centred array sharing six membered rings, in zeolite A the ß cages are separated by four membered rings, in faujasite by six membered rings and in VPI-5 by double four membered rings. This has the effect of opening up the structures

Note, the framework structures of zeolites are frequently drawn showing just the tetrahedral cation centres and connections between them. Hence in Fig. 7.3 each line represents T–T. The oxygen atoms which normally lie slightly off this line are omitted.

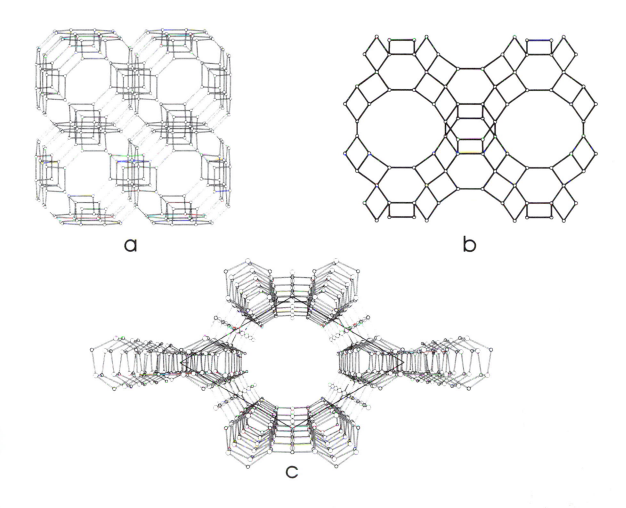

Fig. 7.4 Portions of the framework structures of zeolite A (a), faujasite (b) and VPI-5 (c) showing the manner of ß cage connection and maximum pore size.

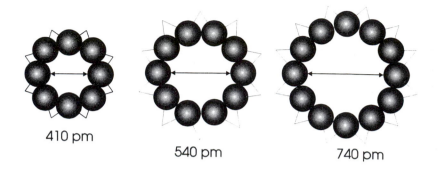

Fig. 7.5 Channel dimensions as a function of ring size. The aperture sizes are determined after allowing for the van der Waals radius of the oxygen atoms around the ring.

410 pm

540 pm

740 pm

and increasing the sizes of the cavities and pores. Hence, in the series sodalite, zeolite-A, ZSM-5, faujasite, VPI-5 the largest channel diameter increases as 260 pm, 410 pm, 540 pm, 740 pm and 1020 pm (2.6-10.2 Å) as the number of tetrahedral units forming the outside of the channel increases from 6 through 8, 10 and 12 to 18, Fig. 7.5.

In materials such as zeolite A and faujasite, which are cubic, the channels run through the crystallites in each of the lattice directions. Other zeolites, for example ZSM-5, Fig. 7.6, are of lower symmetry, in this case orthorhombic, and large channels run parallel to the *b* lattice parameter and are intersected by smaller tunnels.

7.3   The properties of zeolites

Absorption

The open nature of the zeolite structures allows small molecules to be absorbed into their structures; the size and shape of molecules absorbed will depend upon the geometry of the pores. Zeolite A with relatively small pores can absorb molecules such as water and oxygen but larger species, for example ethanol, are barred from entering the cavities. This is illustrated in Fig. 7.7. Note that the type of cation present in the pores can also affect the dimensions of the molecules which can be absorbed. Replacement of sodium

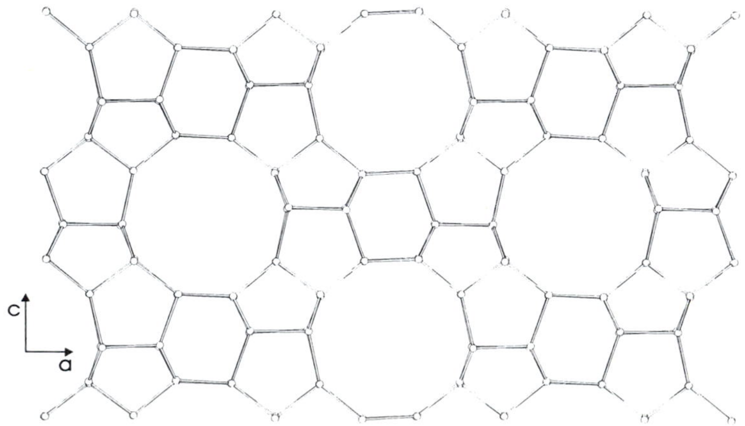

Fig. 7.6 The structure of ZSM-5 viewed down the large pores formed from the 10 membered rings. These large channels are intersected by smaller channels running in the *ab* plane.

c

a

in zeolite A by the smaller calcium increases the effective pore dimension allowing methane to be incorporated.

The ability of an aluminosilicate zeolite to absorb water is related to the nature of the ions forming the framework. With a high number of aluminium atoms in the framework and, therefore, a correspondingly high concentration of charge-compensating cations in the zeolite channels, the structures are highly hydrophillic. Materials, such as sodium-zeolite A, with a silicon to aluminium ratio of 1:1 and a high level of sodium ions in the cavities are widely used to dry gases and solvents. The dehydrating capacity may be regenerated periodically by heating the zeolite and driving off the absorbed water.

Materials such as silicalite, a derivative of ZSM-5 where the framework consists almost totally of SiO_4 tetrahedra, have few cations in the cavities and are hydrophobic. These materials readily absorb non-polar and weakly polar molecules into their cavities.

Ion exchange

The cavity cations in zeolites interact weakly with the framework and may, therefore, undergo ion exchange reactions readily at room temperature. The sodium ions in Na-Zeolite A rapidly exchange with calcium in aqueous solution

$$\text{Na-Zeolite A} + \tfrac{1}{2}Ca^{2+}(aq) \rightarrow Ca_{0.5}\text{-Zeolite A} + Na^{+}(aq) \qquad (7.1)$$

The particular ion exchange characteristics of a zeolite are determined by the sizes of the cages/pores and the coordination environments present within the zeolite. Ion exchange processes are widely used in water softeners and the new 'micro' detergents. In the latter, zeolite A has replaced the polyphosphates previously used, with the environmental advantages of decomposition to soil-like minerals.

Catalysis

The acid derivatives of zeolites, H-zeolite, are excellent catalysts and are widely used industrially. This form of the zeolite may be obtained by direct ion exchange with acids. More often, because many zeolite frameworks are slowly attacked by aqueous acids, they are obtained by exchange with ammonium ions followed by heating to 500°C, driving off ammonia, and leaving the proton

$$\textit{Na-zeolite} \quad \xrightarrow{NH_4^+} \quad \textit{NH}_4\textit{-zeolite} \quad \xrightarrow{500°C} \quad \textit{H-zeolite}$$

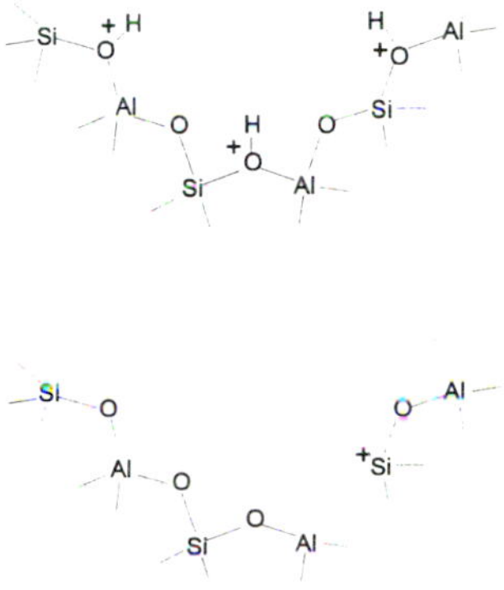

The catalytically active acid zeolite can either be in the Brönsted form with protons attached to the framework tetrahedron (Fig.7.8) or Lewis form obtained by dehydration of the Brönsted acid.

The acid zeolites, once they have absorbed molecules into the cavities, catalyse reactions typical of very strong acids. The major reaction types are

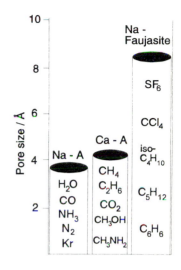

Fig. 7.7 Pore sizes and molecules which may be absorbed for Na-zeolite A, Ca-zeolite A and Na-faujasite.

Fig. 7.8 Brönsted and Lewis acid forms of a zeolite.

dehydrations and rearrangements. A special feature of zeolites which makes them such superb catalysts is their shape selectivity. Zeolites are crystalline materials with controlled channel geometry and fixed environments for the active sites. The shape selectivity may arise in three ways: reactant selectivity, product selectivity and, probably of lesser importance, transition state selectivity. Reactant selectivity arises from the ability of only certain molecules to be absorbed into the zeolite cavities and thus reach the active acid sites. Product selectivity is derived from the fact that only certain products are of the correct dimension to escape rapidly from the zeolite along the channels once they have been formed. This is illustrated for the isomerisation reactions of dimethylbenzenes in Fig.7.9. The 10 ring channels of ZSM-5, with a dimension ≈ 5.4 Å, allow rapid diffusion of 1,4-dimethylbenzene, but not the 1,2 and 1,3 analogues; the diffusivity of the 1,4 derivative is about 10^4 times greater than its isomers. This difference is a result of the effective diameters of these species; the 1,2 and 1,3 placements of the methyl groups make these molecules about 0.6 Å wider than the 1,4 arrangement and diffusion through the 10 rings is hindered. Once a mixture of dimethylbenzenes has entered the ZSM-5 structure, protonation of the benzene ring allows migration of the methyl groups around the ring, equilibrating the isomers. However, the greater mobility of the 1,4 dimethylbenzene arrangement permits this isomer to escape from the zeolite while the other isomers are likely to undergo further protonation and transformation.

Transition state selectivity relies upon the fact that certain intermediates, which are required to be formed during a chemical pathway at the active site, will not fit in the channel/cavity; such a reaction process is barred from occurring and the reaction will proceed along a different route to a different product.

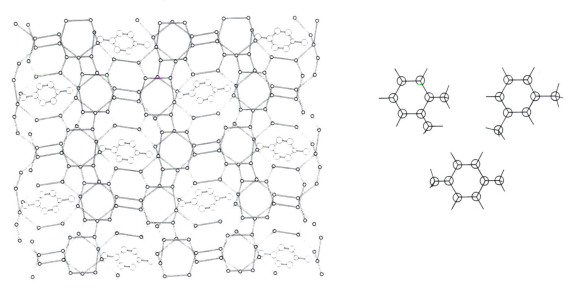

Fig. 7.9 1,4-Dimethylbenzene inside ZSM-5 channels and a comparison of the dimethylbenzene geometries.

7.4 Pillared clays

The largest pore sizes synthesised in zeolites are of the order of 10 Å. In an attempt to increase this further, and allow larger molecules to be absorbed into inorganic structures, chemists have turned to pillared clays.

Clays such as bentonite and montmorillonite have layer structures such as those shown in Fig. 7.10. The layers are constructed of vertex and edge sharing octahedra and tetrahedra. Atoms normally forming the layers are silicon and aluminium plus small mono- and divalent species such as magnesium and lithium. As with zeolites this framework layer has an overall

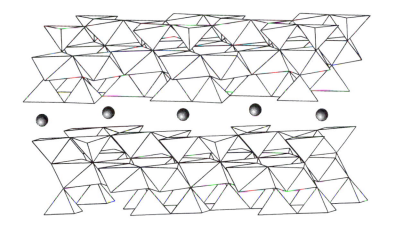

Fig. 7.10 The structure of a clay showing layers formed from linked octahedra and tetrahedra and separated by cations.

negative charge and in clays this is balanced by incorporation of cations, typically alkali metals, between the layers. These inter-layer cations can be readily ion-exchanged.

In the pillaring of clays the species exchanged into the inter-lamellar region is selected for size. Ions such as alkylammonium ions and polynuclear hydroxy-metal ions may replace the alkali metal as shown schematically in Figs. 7.11a/b. The most widely used pillaring species are of the polynuclear hydroxide type and include $Al_{13}O_4(OH)_{28}^{3+}$, $Zr_4(OH)_{16-n}^{n+}$ and $Si_8O_{12}(OH)_8$; the former consists of an AlO_4 tetrahedron surrounded by octahedrally

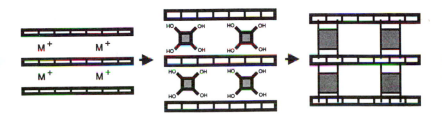

Fig. 7.11 Schematic of pillaring in clays.

(a)   (b)   (c)

coordinated aluminium. This pillaring reaction is readily followed by powder X-ray diffraction; expansion of the inter-lamellar spacing, corresponding to the c lattice parameter, results in a marked shift of the 001 reflection to a lower 2θ value.

Once an ion such as $Al_{13}O_4(OH)_{28}^{3+}$ has been incorporated between the layers, heating the modified clay results in dehydration and linking of the ion to the layers, Fig. 7.11c. The resulting product is a pillared clay which has excellent thermal stability to at least 500°C. The expanded inter-layer region can now absorb large molecules in the same way as zeolites. However, because the distribution of the pillaring ions between the layers is difficult to control, the pillared clay structures are less regular than zeolites. Despite this, pillared clays have been widely surveyed for their potential as catalysts and they act in a similar way to zeolites promoting isomerisation and dehydration.

7.5   Two-dimensional intercalation chemistry

Intercalation between the layers of graphite has been described in Section 3.7 but graphite is only one member of a wide range of compounds which have layered structures and are, therefore, able to incorporate guest species. A number of these materials are summarised in Table 7.1. The most widely studied system has been that of the intercalation compounds of transition metal disulphides.

Intercalation reactions of metal disulphide layer compounds generally involve a redox process where the metal sulphide layer is reduced and the intercalating species oxidised during the reaction. Hence, many species which are readily reduced or can donate electrons can be inserted between the sulphide layers; this includes alkali metals, amines and electron rich transition metal complexes.

Alkali metal insertion in transition metal sulphides

Lithium can be incorporated into TaS_2 over the full range Li_xTaS_2 $0<x<1.0$ occupying octahedrally coordinated sites between the layers; as x increases

Table 7.1 Examples of layer compounds which undergo intercalation reactions

Layer compound	Layer type
Graphite, oxychlorides e.g. FeOCl Transition metal disuphides e.g. TaS_2	Uncharged layers
$LiCoO_2$. Clays: montmorillonite, hectorite. Complex titanates and niobates e.g. $LiNbTiO_5$. Zirconium hydrogenphosphate $Zr(HPO_4)_2.nH_2O$	Negatively charged layers separated by cations or with pendant protons
Hydrotalcites, brucite e.g $Zn_2Fe(OH)_6.(CO_3)_{1/2}.nH_2O$	Positively charge layers separated by anions

the c cell parameter which is perpendicular to the layers increases smoothly from 5.7 Å to 6.2 Å (57-62 pm). For the other alkali metals only certain compositions, e.g. $K_{0.1}TaS_2$, $K_{0.18}TaS_2$ and K_xTaS_2 $0.3 < x < 1.0$, can be prepared as single phases; this behaviour is related to staging, Section 6.6, and probably results from the energy requirement to overcome the van der Waals interactions between layers. The Stage IV compound $K_{0.1}TaS_2$ results from filling of sites between every fourth layer. The reason for the staging behaviour seems to be energetic; partial filling of the sites between all layers would have an energy penalty associated with breaking the van der Waals interactions between *all* layers instead of one in every four as for the staged material.

The alkali metal intercalates of TaS_2 may be hydrated by subsequent exposure to water, forming materials such as $Na_{0.5}(H_2O)_nTaS_2$ where hydrated cations exist between the metal sulphide layers. Other solvents may also be used to generate materials such as $NaDMSO.TaS_2$, $DMSO = (CH_3)_2SO$.

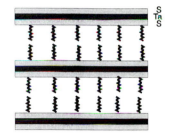

Fig. 7.12 Long chain amine intercalate of a metal disulphide.

Amine intercalation

The intercalation of primary amines into TaS_2 has been achieved for a wide range of chain lengths from methylamine to $C_{18}H_{37}NH_2$, in the latter case forming $(C_{18}H_{37}NH_2)_{0.66}.TaS_2$. In this compound the amine groups exist as a bilayer between the metal sulphide strata, as shown schematically in Fig.7.12. As the number of carbon atoms in the amines increases then the inter-layer separation increases smoothly.

Organometallic intercalation

Intercalation into TaS_2 generally occurs through a redox process and, provided an organometallic species is readily reduced, incorporation of the resulting cation between the metal sulphide layers may be achieved. Cobaltocene, $Co(\eta-C_5H_5)_2$, as a nineteen electron compound is readily oxidised, and forms an intercalate of the stoichiometry $TaS_2.[Co(\eta-C_5H_5)_2]_{0.25}$ by direct reaction in a sealed tube, Fig. 7.13. The eighteen electron compound ferrocene is, however, much more difficult to oxidise and is not incorporated in tantalum disulphide though other, more oxidising, layer compounds such as FeOCl may act as hosts for this compound. Another organometallic which has been inserted into layer disuphides is dibenzenechromium, $Cr(\eta-C_6H_6)_2$.

Fig. 7.13 Cobaltocene intercalated in TaS_2.

7.6 Fast ion conduction in solids

The diffusion of ions through solids is generally very slow, as is apparent in the slow reaction rates of solids (Section 3.1). This is a result of a lack of pathways in solid structures along which ions can migrate; for example for a sodium ion to diffuse through the perfect sodium chloride structure it would have to move through the solid between the sodium and chloride ions on their fixed lattice sites using interstitial positions. Such a migration would have a high energy penalty in disrupting the lattice. One possible pathway for ion

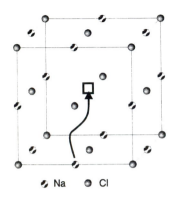

Na   **Cl**

Fig. 7.14 An ion conduction pathway in NaCl using a Schottky defect. Once one ion has migrated to the empty site the vacancy left behind may be filled by a different sodium ion. Ion conduction thus occurs by a series of hops.

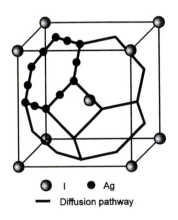

I   **Ag**
— **Diffusion pathway**

Fig. 7.15 The iodide ion framework of α-AgI and the regions occupied by silver ions.

diffusion through a material with the sodium chloride structure would be by making use of defects in the structure such as Schottky defects (Section 6.2). Sodium ions may diffuse through the structure using the vacancies, Fig. 7.14. As the number of these defects increases so the rate of ion diffusion increases and just below its melting point sodium chloride is a reasonable ion conductor, with a conductivity of 10^{-3} $\Omega^{-1}cm^{-1}$. Similar behaviour may be found with anions and ZrO_2 is a good conductor of oxide ions above 600°C.

The diffusion of ions through solids is thus facilitated by the presence of suitable vacant sites to which a diffusing ion may migrate. In some materials the presence of a large number of suitable sites within the structure for ion migration, even at low temperatures, allows facile diffusion of ions and these materials are known as fast ion conductors. Their conductivities reach values in the range 10^{-3}–10^1 $\Omega^{-1}cm^{-1}$ comparable to ionic solutions. Typical structural and chemical features exhibited by fast ion conductors are vacant sites to which ions may migrate and small cations with a low charge; these interact weakly, in electrostatic terms, with the rest of the structure and form low coordination number environments. Some of the best ion conductors are, therefore, based on silver and lithium.

One example of a fast ion conductor is α-AgI. At 146°C silver iodide adopts a structure based on a body centred array of iodide ions shown in Fig.7.15. The silver atoms are able to occupy a range of sites throughout this iodide matrix close to the face centres and movement between these positions is facile.

Some of the best fast ion conductors consist of a stable framework constructed from octahedral or tetrahedral metal oxo species which form channels or pathways along which cations can migrate. These channels are somewhat smaller than those in zeolites as only small ions, typically alkali metals and silver, need pass along them. Examples of fast ion conductors of this type are NASICON (<u>Na</u> <u>S</u>uper<u>I</u>onic <u>CON</u>ductor) which is $Na_3Zr_2PSi_2O_{12}$, and $Li_{14}ZnGe_4O_{16}$ (a lithium ion conductor) and a number of derivatives of the two dimensional material ß-alumina.

The NASICON structure, Fig. 7.16, has channels, formed from ZrO_6 octahedra and SiO_4 tetrahedra, containing the sodium ions. There are more possible sites for sodium to occupy in the channels than sodium ions present and by hopping between these positions sodium ion diffusion along the channels is possible.

The ß-aluminas are a class of compounds of the general formula $M_2O.nAl_2O_3$ where $n=8$–11 and M is a very wide range of cations which includes Na^+, K^+, Rb^+, Ag^+, NH_4^+, H_3O^+, NO^+, Mg^+ and Ln^{3+}. The structure, Fig. 7.17, consists of blocks which have structures similar to that of the spinels separated by oxide ion deficient regions in which reside the cations, M. The sections containing the M cations provide a large number of possible sites and migration of these ions is rapid within these layers. At 300°C the conductivity of sodium ß-alumina is about 0.1 $\Omega^{-1}cm^{-1}$.

Conduction Channels

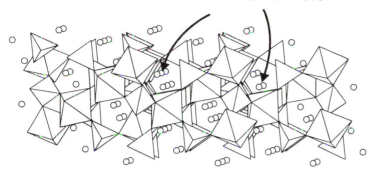

Fig. 7.16 The structure of NASICON showing the channels containing the mobile sodium ions.

Applications of fast ion conductors include batteries, where the solid cation conductor acts as the electrolyte, fuel cells and oxygen gas sensors.

7.7 Problems

7.1 The powder diffraction pattern collected from ZSM-5 (λ=1.54 Å, 154 pm) has reflections at 7.868°, 8.767° and 22.110°, which can be indexed as the 101, 200 and 050 reflections respectively. Calculate the orthorhombic unit cell parameters. Using Fig. 7.6, estimate the size of the 10 membered ring apertures in ZSM-5.

7.2 The 001 reflection collected from $TaS_2.C_{18}H_{35}NH_2$ (λ=1.54 Å, 154 pm) is at $2\theta = 1.57°$ whilst the corresponding reflection in TaS_2 is at 17.58°. Calculate a length for this amine chain assuming that they are perpendicular to the sulphide layers. How would the position of the 001 reflection change if the amine groups were tilted at an angle relative to the TaS_2 layers?

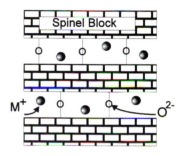

Fig. 7.17 Schematic of the ß-alumina structure.

8 Some recent developments in inorganic materials chemistry

Many of the recent advances in inorganic chemistry have come in the area of new materials. The chemistry of high temperature superconducting phases, in terms of the synthesis and structure of these compounds, lies within the area of this text. More recently the structural chemistry of C_{60} in its ionic compounds is also relevant, though the chemistry of this species is developing rapidly in many areas

8.1 High temperature superconductors

Superconductivity was first observed by K.Onnes in 1911; on cooling a sample of mercury below 4.2 K the resistivity of the metal suddenly decreased to an immeasurably small value, Fig. 8.1. Since then many metals and alloys have been found which have zero resistance below a certain critical temperature, T_c. The progress in terms of raising T_c, in metals and alloys since 1911, is summarised in Fig. 8.2.

Above T_c superconducting materials show a finite resistance in common with other compounds. This electrical resistance results from the interaction between lattice vibrations and the conduction electrons as they move through the structure. This interaction or scattering of the conduction electrons strengthens as the atomic vibrations increase in amplitude, that is, as the temperature of the system increases. Hence, the resistance of metals rises as they are heated. In superconducting metals and alloys it is believed that the electrons move through the lattice in a *concerted* fashion with the lattice vibrations, resulting in no electron scattering and zero resistance.

Not all superconducting materials are pure elemental metals and their alloys. By 1985, zero resistance at low temperatures had been observed in many compounds, for example metal oxides Li_2TiO_4, T_c = 13.7 K (spinel type structure), sulphides $PbMo_6S_8$, T_c= 15.2 K and organic metals formed from charge transfer compounds (up to 13 K); all of these systems exhibit metal-like behaviour in terms of their conductivity above their T_c. In general though the critical temperatures of these compound types are fairly low, mostly below 10 K.

In 1986, J.G.Bednorz and K.A.Müller published work describing the observation of anomalously low resistance in a mixed metal oxide system containing barium, lanthanum and copper. The compound was later shown to be $La_{2-x}Ba_xCuO_4$, with a critical temperature of 35 K, and Bednorz and Müller

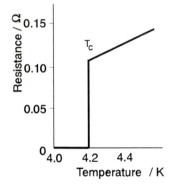

Fig. 8.1 Resistiance of a sample of mercury as a function of temperature.

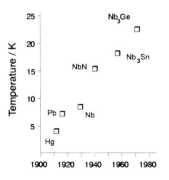

Fig. 8.2 Increases in T_c since 1911.

were awarded the Nobel Prize for their discovery. Since then progress in developing new inorganic materials with higher T_c's has been extremely rapid. These new materials are all complex copper oxides and, at the time of writing, the critical temperature in bulk materials has been increased to 134 K with reports of even higher T_c's under high pressures and in thin films. The structures and chemistry of the most important of these superconducting phases are discussed in the following sections.

$La_{2-x}M_xCuO_4$, M=Ba, Sr

La_2CuO_4 has a structure similar to that of K_2NiF_2 described in Section 4.7 though the elongated CuO_6 octahedra are tilted relative to the 001 plane of the unit cell. Partial substitution of the trivalent lanthanum by divalent barium, whilst maintaining the oxygen content, produces $La_{1.8}Ba_{0.2}CuO_4$, Fig. 8.3, which has the perfect K_2NiF_4 structure. This material has a T_c of 35 K, the strontium doped analogue, $La_{1.85}Sr_{0.15}CuO_4$, has a T_c of 40 K.

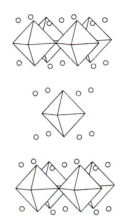

Fig. 8.3 The structure of $La_{1.8}Ba_{0.2}CuO_4$.

$YBa_2Cu_3O_{7-\delta}$, $0 < \delta < 1$

The non-stoichiometric $YBa_2Cu_3O_{7-\delta}$ system has been the most widely studied of the superconducting materials. This is a result of its ease of synthesis and moderately high critical temperature, 93 K, which lies reasonably above the boiling point of nitrogen (77 K). One complicating factor in this system has been the control of the oxygen stoichiometry which is important in determining the critical temperature. The best superconducting properties are observed when $\delta \approx 0$, $YBa_2Cu_3O_{6.9-7.0}$ with $T_c = 93$ K, as δ increases then T_c falls rapidly as shown in Fig. 8.4. In the synthesis of $YBa_2Cu_3O_7$ it is, therefore, important to control the oxygen stoichiometry which requires a two stage reaction process. The direct high temperature reaction of $BaCO_3$, Y_2O_3 and CuO at 940°C over a period of days followed by quenching to room

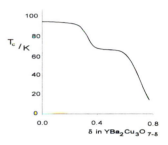

Fig. 8.4 Variation of T_c in $YBa_2Cu_3O_{7-\delta}$ with δ.

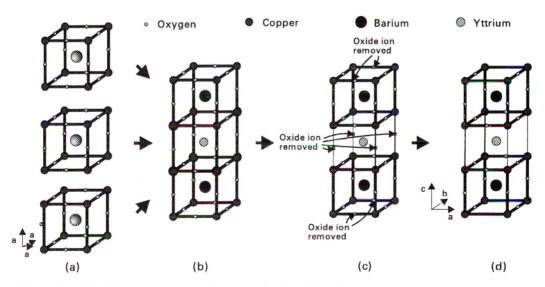

Fig. 8.5 Derivation of the $YBa_2Cu_3O_7$ structure from a simple perovskite cube.

temperature gives rise to material deficient in oxygen with a stoichiometry of about $YBa_2Cu_3O_{6.7}$. A material with a similar stoichiometry may be prepared using the gel methods described in Chapter 3. In order to produce the optimum oxygen content these oxygen deficient materials must be annealed in pure oxygen, at temperatures of 400–500°C, followed by slow cooling to room temperature. Under these conditions the material takes up additional oxygen and δ falls close to zero.

The structure of $YBa_2Cu_3O_7$ is derived from that of perovskite. Stacking three perovskite unit cells, of the stoichiometry ABO_3, directly on top of each other, as in Fig. 8.5a, gives rise to a material of stoichiometry $A_3B_3O_9$ (3 × ABO_3). If the A type cations are replaced by 2 × Ba and 1 × Y in the sequence Ba-Y-Ba-Ba-Y-Ba-Ba-Y in the tripled perovskite, and the B cations are designated as copper then the compound stoichiometry, shown in Fig. 8.5b, becomes $YBa_2Cu_3O_9$. Oxygen is then removed from the (00½) site in the structure adjacent to yttrium and the site in the basal plane at (½00) giving the $YBa_2Cu_3O_7$ structure shown in Fig. 8.5d. The cell parameters are based on those of the original perovskite cube with $a \approx 390$ pm in $YBa_2Cu_3O_7$, c is approximately treble a and, because of the selective removal of oxygen from the sites in the basal plane, a and b are different. The crystal system of $YBa_2Cu_3O_7$ is thus orthorhombic with the lattice constants $a = 385$ pm, $b = 388$ pm and $c = 1140$ pm.

Consideration of Fig. 8.5d shows that $YBa_2Cu_3O_7$ contains copper in two different coordination environments, those with fractional coordinates near (0,0,⅓) in square pyramidal sites and those at the origin in square planar sites. A polyhedral description of the structure using the copper coordination geometry, Fig. 8.6, makes this clearer. The linking of these sites leads to them being referred to as forming chains and planes.

Fig. 8.6 The structure of $YBa_2Cu_3O_7$ showing the copper oxygen polyhedra. Square planar sites link together forming one dimensional chains along the a axis whilst square pyramidal sites link together to form a two-dimensional layer.

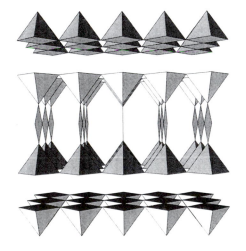

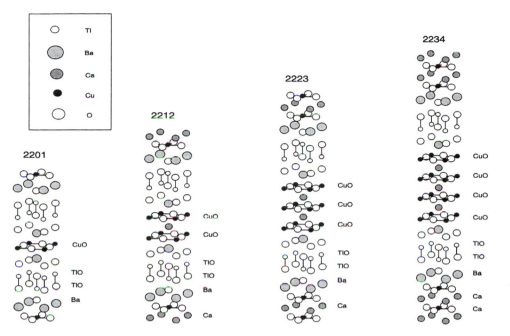

Fig. 8.8 The structures of $Tl_2Ba_2Ca_{n-1}Cu_nO_{4n+2}$ with n = 1, 2, 3 and 4.

The oxygen stoichiometry in $YBa_2Cu_3O_{7-\delta}$ may be varied with $0<\delta<1$ and in terms of the structure this is achieved by removing oxygen from the other basal plane site, (0,½,0). Complete removal of this oxygen yields $YBa_2Cu_3O_6$, Fig. 8.7. The coordination of the 'chain sites' drops to linear, typical of Cu^+.

$Tl_2Ba_2Ca_{n-1}Cu_nO_{4n+2}$, n = 1, 2, 3, 4

Most of the highest T_c's observed in bulk materials have been found in the system of thallium barium calcium cuprates. A series of materials with differing numbers of CuO_2 layers have been prepared in the homologous series $Tl_2Ba_2Ca_{n-1}Cu_nO_{4n+2}$ with n = 1, 2, 3, 4 and the structures of these materials are shown in Fig. 8.8. The structure of $Tl_2Ba_2CaCu_2O_8$ (2212) consists of two layers of CuO_2 square nets separated by a double thallium–oxygen layer, Tl_2O_2. The copper–calcium–oxygen layer may be considered as being derived from a perovskite block, with oxygen vacancies, whilst the Tl_2O_2 layer has an arrangement of ions similar to that of a portion of the rocksalt (NaCl) structure. As n increases in the formula $Tl_2Ba_2Ca_{n-1}Cu_nO_{4n+2}$ then additional CuO_2 layers are incorporated into the structure, thus $Tl_2Ba_2Ca_2Cu_3O_{10}$ and $Tl_2Ba_2Ca_3Cu_4O_{12}$ have three and four cuprate layers respectively. All these materials have tetragonal unit cells with $a\approx3.8$ Å (380 pm) but different c parameters; as the number of CuO_2 layers increases so c increases.

A series of materials with only a single thallium–oxygen layer separating the CuO_2 sheets is also known. For example, the structure of $TlBa_2Ca_2Cu_3O_9$

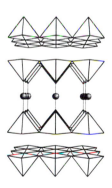

Fig. 8.7 The structure of $YBa_2Cu_3O_6$.

may be derived from that of $Tl_2Ba_2Ca_2Cu_3O_{10}$ by removal of one of the TlO layers.

The critical temperature behaviour of the $Tl_2Ba_2Ca_{n-1}Cu_nO_{4n+2}$ system is very complex and depends upon the exact compound stoichiometry, synthesis conditions and subsequent annealing treatments in different atmospheres. One trend is apparent, T_c rises as n increases from 1 to 3, 80 K in $Tl_2Ba_2CuO_6$, 105 K in $Tl_2Ba_2CaCu_2O_8$ and 125 K in $Tl_2Ba_2Ca_2Cu_3O_{10}$; for the next member of the series, $Tl_2Ba_2Ca_3Cu_4O_{12}$, however, T_c falls back to around 115 K.

$Bi_2Sr_2Ca_{n-1}Cu_nO_{4n+2}$, $n=1, 2, 3$ and $HgBa_2Ca_2Cu_3O_{10}$

A series of compounds analogous to the thallium–barium–calcium–copper oxide superconductors, but containing Bi_2O_2 rather than Tl_2O_2 layers, has been synthesised. As the Bi–O distance is slightly larger than the Tl–O bond length, the Bi_2O_2 layers fit poorly between the cuprate sheets causing some structural distortions; the compounds adopt orthorhombic unit cells and critical temperatures are somewhat lower than the thallium analogues.

The compound $HgBa_2Ca_2Cu_3O_{10}$ was reported in early 1993 with a T_c of 134 K. Its structure is similar to that of $TlBa_2Ca_2Cu_3O_9$, but with a single HgO layer separating three CuO_2 planes. The compound is difficult to make as a pure phase, but illustrates that further upward progress in T_c is still possible; indeed under pressure the Tc can be raised to 160 K.

Superconducting cuprates — general points

Consideration of the several superconducting systems described above shows two features in common. They all contain planes of the stoichiometry CuO_2 formed by linking CuO_4 square planes (elongated CuO_6 octahedra in $La_{1.8}Ba_{0.2}CuO_4$) at all vertices, Fig. 8.9, and they all have an average copper oxidation state in excess of 2+. These CuO_2 sheets are separated by 'charge reservoir' layers, e.g. CuO chains in $YBa_2Cu_3O_7$ and the Tl_2O_2 rocksalt layers in $Tl_2Ba_2Ca_{n-1}Cu_nO_{4n+2}$, which seem to ensure the correct charge or electron concentration on the superconducting sheets.

Fig. 8.9 Schematic of the structural elements present in a high temperature superconducting cuprate.

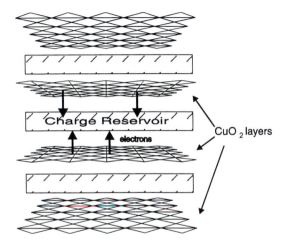

Consideration of the numerous superconducting cuprates which have been synthesised in the last five years shows these features to be recurring. The search for new superconducting materials is thus concentrating on compounds which show these structural characteristics.

8.2   Fullerenes and Fullerides

The chemistry of C_{60}, Fig. 8.10, has crossed many of the conventional borders of chemistry and physics with interest in this molecule covering many areas. These fields of study include the organic and organometallic chemistry of C_{60} as a ligand. The structural chemistry of the solid fullerene C_{60} and the $M_n C_{60}$ fulleride derivatives are covered here.

C_{60}

The first crystals of C_{60} were grown from benzene solutions and contained solvent molecules. With the correct purification methods, using sublimation to eliminate the solvent molecules, pure C_{60} crystals may be grown. The compound adopts a face centred cubic array of the C_{60} molecules shown in Fig. 8.11, which is equivalent to the close packing of C_{60} molecules. At room temperature the thermal energy available to the molecules allows them to rotate freely on their lattice sites and they can be considered to be spherical. The powder X-ray diffraction pattern of polycrystalline C_{60}, shown in Fig. 8.12, is typical of the face centred cubic lattice though due to the high symmetry of the spherical molecules, reflections with Miller indices $h00$ have zero intensity and are not seen. The lattice parameter of 14.17 Å obtained from the powder data yields a contact distance of $14.17/\sqrt{2}$ or 10.02 Å for the C_{60} spheres, which is made up of the diameter of a C_{60} molecule, 7.06 Å, plus a van der Waals distance of 2.96 Å between the molecules. This van der Waals distance is slightly smaller than the value in graphite, corresponding to the inter-layer separation, of 3.35 Å.

Fig. 8.10 The structure of a C_{60} molecule.

Note that one of the most efficient ways of packing spheres, in terms of filling space, is cubic close packing. Given the shape of the C_{60} molecule it is not surprising that they adopt the face centre cubic lattice.

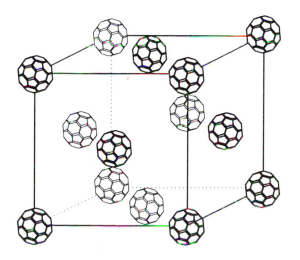

Fig. 8.11 The unit cell of solid C_{60}.

Fig. 8.12 The powder X-ray diffraction pattern of solid C_{60}.

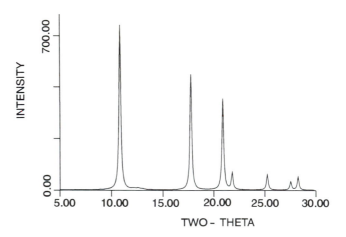

Fig. 8.13 The structure of K_3C_{60}.

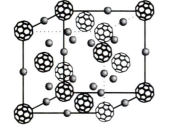

The intercalation compounds of C_{60}

Reaction of C_{60} with potassium vapour and other alkali metals forms compounds of the type A_xC_{60}, the stoichiometry of the product depending on the availability of the reactant. With excess alkali metal, compounds of the stoichiometry A_6C_{60}, A = K, Rb, Cs are formed. The structure of K_6C_{60} is body centred cubic; C_{60} molecules occupy sites at the cell corners and body centre whilst the potassium atoms fill a number of sites near the centres of the faces. Of most interest of the other stoichiometries are the compounds A_3C_{60} which have been found to be superconducting at reasonably high temperatures. The powder X-ray diffraction pattern of K_3C_{60} is similar to that of the parent C_{60} in Fig. 8.12. The peaks are shifted to lower 2θ values indicating an expansion of the C_{60} face centred cubic lattice. The stoichiometry K_3C_{60} is obtained by filling all the tetrahedral and all the octahedral holes in the close packed C_{60} lattice and is shown in Fig. 8.13.

K_3C_{60} becomes superconducting at 19 K, though gradual replacement of potassium by the larger alkali metal ions raises T_c, so that Rb_3C_{60} has a T_c of 29 K and $CsRb_2C_{60}$ a value of 33 K.

8.3   Problems

8.1 The compound $La_2SrCu_2O_6$ has the same structure as $Sr_3Fe_2O_6$, shown in Fig. 6.15, but with lanthanum replacing some of the strontium. Would you expect this compound to be a high temperature superconductor?

8.2. The 111 reflection ($\lambda=1.54$ Å, 154 pm) of K_3C_{60} is at 10.75°. Calculate the shortest potassium–potassium distance in this compound using the description of the structure given above.

Answers to problems

1.1. F, F, P, I.

1.2. Draw the unit cell in projection down b. The short diagonal of the parallelogram becomes the face of the face centred lattice.

1.3. Orthorhombic; $d_{101}=d_{10-1} = 2.40$ Å (240 pm). Monoclinic; $d_{101}= 1.956$ Å (195.6 pm), $d_{10-1}= 2.752$ Å (275.2 pm).

2.1. $a = 12.358$ Å (1235.8 pm), lattice type P. Maximum aperture, wall to wall is about 9 Å (90 pm).

2.2. F lattice 33.43 (111), 38.70 (200), 56.03 (220), 66.84 (311) and 70.23 (222); I lattice 34.83 (110), 50.09 (200), 62.46 (211) and 73.55 (220). The peaks would move to lower 2θ values as the metal expands.

2.3. Peak splits giving 3 reflections. 310,301,103. with similar intensities.

3.1. Use precursors/precipitation methods.

3.2. The metal–oxygen framework is determined in the high temperature reaction. $KNbTiO_5$ has a different structure from $LiNbTiO_5$; the ion exchange of Li for K leaves the $[NbTiO_5]^-$ framework essentially unchanged.

3.3. The liquid cools until, just below T_2, the liquidus is reached; solid AB precipitates and the liquid becomes richer in B. This process continues with further cooling until the eutectic is reached and the remaining liquid solidifies to give further AB + B.

4.1. $t=1.037$. Similar behaviour to $BaTiO_3$.

4.2. $FeCr_2O_4$ normal spinel Cr^{3+} has very strong preference for octahedral sites. $NiGa_2O_4$ inverse spinel Ni^{2+} has a preference for octahedral sites.

5.1. $Li_{0.3}ReO_3$ is metallic; the π-level has 0.3 extra electrons per rhenium, as compared with ReO_3, but is still only partially filled.

5.2. The d_{xy} orbital points along the M–M direction. In TiO_2 a band formed from the overlap of these orbitals is empty and TiO_2 is an insulator. In VO_2 it is partially filled and VO_2 demonstrates metallic properties.

6.1. ≈ 0.6 Å, hence need for vacancies on normal sites.

6.2. The a parameter decreases *smoothly* (a non-stoichiometric system) from $Sr_3Fe_2O_6$ to $Sr_3Fe_2O_7$ as the Fe^{3+} is oxidised to the smaller Fe^{4+}. The c parameter initially gets larger as the FeO_5 square pyramids are pushed apart to make room for the oxide ion then it also decreases smoothly.

7.1. $a=20.15$ Å, $b=20.08$ Å and $c =13.51$ Å. The wall to wall distance is ≈ 8.5 Å taking no account of the radius of the oxygen atom.

7.2. The interlayer spacing is 56.2 Å in the amine derivative and 5.03 Å in TaS_2. Hence the amine chain length must be ≈ 25 Å. If the amine chains were tilted the interlayer spacing would be reduced and the 001 reflection would move to larger 2θ values.

8.1. $La_2SrCu_2O_6$ is not a superconductor. The structure contains CuO_2 planes but the copper oxidation state is 2.0 exactly.

8.2. 6.17 Å (617 pm)

Bibliography

1. *Inorganic solids.* Adams, D. M. (1974). Wiley, New York.

2. *Modern powder diffraction: reviews in mineralogy volume 20.* Bish, D. L. and Post, J. E. (ed.) (1989). The Mineralogical Society of America, Washington DC.

3. *Inorganic materials.* Bruce, D. W. and O'Hare, D. (ed.) (1992). Wiley, New York.

4. *Solid state chemistry: techniques.* Cheetham, A. K. and Day, P. (ed.) (1987). Oxford University Press.

5. *Solid state chemistry: compounds.* Cheetham, A. K. and Day, P. (ed.) (1992). Oxford University Press.

6. *Electronic structure and chemistry of solids.* Cox, P. A. (1987). Oxford University Press.

7. *Transition metal oxides.* Cox, P. A. (1993). Oxford University Press.

8. *An introduction to zeolite molecular sieves.* Dyer, A. (1988). Wiley, New York.

9. *Ionic crystals, defects and non-stoichiometry.* Greenwood, N. N. (1968). Butterworths, London.

10. *Inorganic structural chemistry.* 2nd edition. Müller, U. (1992). Wiley, New York.

11. *Solid state chemistry: an introduction.* Smart, L. and Moore, E. (1992). Chapman and Hall, London.

12. *Structural inorganic chemistry.* 5th edition. Wells, A. F. (1984). Oxford University Press.

13. *Solid state chemistry and its applications.* West, A. R. (1984). Wiley, New York.

Index

Where chemical formula are indexed they are provided at the end of each letter section.

DATE DUE
